C0-AVZ-622

Greening India's Growth

India's sustained and rapid economic growth offers an opportunity to lift millions out of poverty. But this may come at a steep cost to the nation's environment and natural resources. This insightful book analyzes India's growth from an economic perspective and assesses whether India can grow in a "green" and sustainable manner. Three key issues are addressed.

The first is the physical and monetary costs and losses of environmental health and natural resources driven by economic growth. The authors undertake a monetary valuation and quantification of environmental damage, using techniques that have been developed to better understand and quantify preferences and values of individuals and communities in the context of environmental quality, conservation of natural resources, and environmental health risks. The second part estimates the value of ecosystem services from the major biomes in India using state-of-the art methods with a view to preserving them for the future. The third section provides a menu of policy instruments to explore trade-offs between economic growth and environmental sustainability using a computable general equilibrium approach with particular attention to air pollution.

The conclusions focus on the way forward in terms of policies, measures, and instruments as India seeks to balance the twin challenges of maintaining economic prosperity and simultaneously managing its environmental resources.

Muthukumara S. Mani is a senior environmental economist in the Disaster Risk and Climate Change Unit, South Asia Sustainable Development Department, with the World Bank in Washington, DC.

"*Greening India's Growth* is timely, very relevant and provides a framework for a pattern of development that is designed to bring about economic development in a manner that imposes the least impact on the environment, as well as on resource intensity, and therefore embodies fully the principle of sustainability. I welcome the book's contribution to the debate between economic growth and environmental sustainability in the Indian context.

It rightly argues that the way to resolve the conflict between environment and development is to make the tradeoffs explicit. By putting numbers to the environmental costs of our growth process, the book highlights why, 'Grow now, clean up later' is no longer an option for India. Such analysis must be mainstreamed into our policies and plans.

Like in many other countries, the debate over growth versus environment is also active in India. This book makes a significant contribution to this debate by providing analytical insights on how failure to act now could constrain long-term productivity and hence growth prospects. It also highlights the needs to compute green Gross Domestic Product (green GDP) as an index of economic growth with relevant environmental costs and services factored in."

R. K. Pachauri, Director-General, The Energy and Resource Institute, India and Chairperson of the Intergovernmental Panel on Climate Change (which was awarded the Nobel Peace Prize in 2007)

"*Greening India's Growth* makes a significant contribution to the 'growth versus environment' debate by suggesting that there are low-cost policy options that could significantly bring down environmental damage without compromising long-term growth objectives. By linking sustainability with public health and livelihood issues, the book highlights the need for Green National Accounting so that environmental costs and services are factored into the growth process."

Jairam Ramesh, Minister of Rural Development and Minister of Drinking Water and Sanitation, and Former Minister of State at the Ministry of Environment and Forests, India

"Like in many other countries, the debate over growth versus environment is also active in India. *Greening India's Growth* makes an important contribution to this debate by suggesting that there are low-cost options to reducing environmental damage that are not only affordable in the long-term but would also be offset by the significant health and productivity benefits."

Onno Ruhl, India Country Director, World Bank

Greening India's Growth

Costs, Valuations, and Trade-offs

Edited by Muthukumara S. Mani

First published 2014
by Routledge
2 Park Square, Milton Park, Abingdon, Oxon OX14 4RN

and by Routledge
711 Third Avenue, New York, NY 10017

Routledge is an imprint of the Taylor & Francis Group, an informa business

British Library Cataloguing-in-Publication Data
A catalogue record for this book is available from the British Library

Library of Congress Cataloging-in-Publication Data

Mani, Muthukumara, 1964–
Greening India's growth : costs, valuations and trade-offs / edited by Muthukumara Mani.
pages cm
Includes bibliographical references and index.
1. Sustainable development—India. 2. Economic development—Environmental aspects—India. I. Title.
HC440.E5M37 2014
338.954'07—dc23
2013025148

ISBN: 978-0-415-71935-3 (hbk)
ISBN: 978-1-315-86714-4 (ebk)

Typeset in Sabon
by Apex CoVantage, LLC

Printed and bound in the United States of America by
Edwards Brothers Malloy

Contents

Foreword

Strong investment, reflecting rising productivity, healthy corporate profits, and robust exports, has fueled economic growth exceeding 7 percent a year in India for almost a decade. That growth in turn has increased employment opportunities and allowed millions to emerge from poverty.

As a result, India has emerged as a major power with an economy ($4.7 trillion) that in 2012 became the world's third-largest (in purchasing power terms), surpassing Japan and now positioned behind only China and the United States. Its trade in goods and services is close to a trillion dollars and is expected to double every seven years. Even with the recent slowdown some economists think India will grow faster than any other large country over the next 25 years. But does growth—so essential for development—have to come at the price of worsened air quality and other environmental impacts?

Surprisingly, few assessments consider the environmental sustainability of growth or the impacts of ecosystem degradation and natural disasters on development outcomes. Economic expansion will be accompanied by rising demands on already scarce and often degrading natural resources (soils, energy, watersheds, and forests) and a growing carbon and pollution footprint that negatively affects human health and growth prospects. Climate change and increasing frequency and intensity of extreme events are expected to further exacerbate these already serious public health problems.

What is sustainable growth and sustainable development for India? Why is it important? How do we measure it? How do we ensure it? Where might the balance lie between rising gross domestic product (GDP) and declining environmental assets? This study seeks to address these fundamental questions that would determine the sustainability of India's current development trajectory. It provides an assessment of environmental changes and development impacts that can help define priorities and environmental management strategies.

The study:

- provides estimates of social and financial costs of current environmental damage in India;
- provides an economic valuation of ecosystem services across the country using quantified estimates of landscape types; and
- assesses trade-offs between economic growth and environmental sustainability for India using an economy-wide modeling approach.

To achieve these objectives, the study has been divided into three specific topical areas:

1. The first topical area, "How Much Does It Cost?," looks at the annual physical and monetary losses in environmental health and natural resources driven to some extent by economic growth. It undertakes a monetary valuation and quantification of environmental damage, using new techniques and methodologies that have been developed in recent decades to better understand and quantify preferences and values of individuals and communities in the context of environmental quality, conservation of natural resources, and environmental health risks.
2. The second topical area, "How to Value?," estimates the value of ecosystem services from the major biomes in India using state-of-the-art methods with a view to preserving them for the future.
3. The third topical area, "What Are the Trade-Offs?," provides a menu of policy instruments to explore trade-offs between economic growth and environmental sustainability using a computable general equilibrium approach, with particular attention to air pollution.

Given the growing population, urbanization, and the drive toward greater economic well-being, there is little or no doubt that the environmental challenges will continue to increase steadily. That being the case, there is no doubt that substantive strengthening of the policy and institutional framework will be required in order to manage the greater challenges. The conclusions of this study focus on the way forward in terms of policies, measures, and instruments as India balances the twin challenges of maintaining economic prosperity and managing its environmental resources.

Greening India's growth requires strategies that will break the pattern of environmental degradation and natural resource depletion that are too often the consequence of economic growth and that will avoid locking the economy into unsustainable patterns.

This study was conducted with wide consultation with the government, nongovernmental organizations, and research and academic institutes in India, and their contribution is gratefully acknowledged.

Philippe Le Houérou
Vice President, South Asia

Contributors and Acknowledgments

The study was led by Muthukumara S. Mani (task team leader) of the South Asia Sustainable Development Department of the World Bank, and the core team included Sonia Chand Sandhu, Gaurav Joshi, Anil Markandya, Sebnem Sahin, Elena Strukova, Aarsi Sagar, Vaideeswaran Sankaran, Bela Varma, and Marie Elvie. The team gratefully acknowledges the contributions of Dan Biller, Charles Cormier, Giovanna Prennushi, Herbert Acquay, and Michael Toman, who carefully reviewed the work and provided expert guidance to the team at crucial stages.

John Henry Stein, sector director of the South Asia Sustainable Development Department, and Onno Ruhl, country director for India, guided the overall effort.

The study was conducted with wide consultation with the Ministry of Environment Forests, nongovernmental organizations, and research and academic institutes in India, and their contribution is gratefully acknowledged.

Financial assistance was also provided by the UK Department for International Development, the Trust Fund for Environmentally and Socially Sustainable Development of the governments of Finland and Norway, and the Trust Fund for Bank-Netherlands Partnership Program.

Chapter 1

Growth versus Environment Debate

Economic policies designed to promote growth have been implemented without considering their full environmental consequences, presumably on the assumption that these consequences would either take care of themselves or could be dealt with separately. These are serious consequences, and it has become clear today that economic development must be environmentally sustainable.

Dr. Manmohan Singh, Prime Minister of India
International Workshop on Green National Accounting for India, in New Delhi,
April 2, 2013

Buoyant economic growth exceeding 7 percent over the last decade has raised the tantalizing prospect that India could eliminate extensive poverty within a generation. Growth has been fueled by a strong momentum in investment reflecting rising productivity, healthy corporate profits, robust exports, and high business confidence. As a result of this unprecedented economic expansion, India has become, in purchasing power parity terms, the world's third-largest economy, behind the United States and China. Even with the recent slowdown, some economists think India will grow faster than any other large country over the next 25 years.[1] An implicit assumption behind these optimistic predictions is that the conditions that have made the current rapid growth possible will prevail in the future.[2]

Surprisingly, none of these assessments consider the environmental sustainability of growth or the impacts of ecosystem degradation on development outcomes. Economic expansion will be accompanied by rising demands on already scarce and often degraded natural resources (soils, fossil fuels, water, and forests) and a growing pollution footprint that impacts negatively on human health and growth prospects.

Will the quest for high growth result in unacceptable environmental loss that will ultimately impede poverty alleviation? Should India grow now and clean up later? Where might the balance lie between rising gross domestic product (GDP) and declining environmental assets? This study seeks to address these fundamental questions, which are important because they concern the sustainability of India's current development trajectory. This study provides an assessment of environmental changes and development impacts that can help define priorities and environmental management strategies.

1.1 The Environmental Challenges of Rapid Growth

Economic growth is universally recognized as a prerequisite for development. In India economic growth added eight million jobs every year between the early 1990s and 2004–2005 and allowed millions to emerge from poverty.[3] The national poverty ratio halved over this period (National Institute of Rural Development, 2003), and by some estimates 300 million have joined the middle class with an income of at least US$7,000 per year.[4]

India's remarkable growth record has been clouded by a degrading environment and growing scarcity of natural resources. Mirroring the size and diversity of the nation's economy, environmental risks are wide ranging and are driven by both poverty and prosperity. Much of the burden of growth and development is falling upon the country's natural assets and its people.

In a recent survey of 132 countries whose environments were surveyed, India ranked 125th overall and last in the "Air Pollution (effects on human health)" category.[5] The study concluded that India has the worst air pollution in the entire world, beating China, Pakistan, Nepal, and Bangladesh. The World Health Organization's (WHO) recent Global Burden of Disease assessment estimates that outdoor air pollution causes 620,000 premature deaths per year in India, a sixfold increase since 2000. The main cause is growing levels of particulate emissions (PM10) from transport and power plants. Also, according to the WHO, across the G-20 economies, 13 of the 20 most polluted cities are in India, and more than 50 percent of the sites studied across India had critical levels of PM10 pollution. A recent rapid survey by the Delhi-based

Centre for Science and Environment revealed that almost 75 percent of respondents considered air pollution to be a major cause of concern and responsible for respiratory illnesses.

Simultaneously, poverty remains both a cause and a consequence of resource degradation: agricultural yields are lower on degraded land, and when forests are depleted, livelihood resources decline. To subsist, the poor are compelled to mine and overuse the limited resources available to them, creating a downward spiral of impoverishment and environmental degradation. The problem is most visible in the lagging regions of India, where rural poverty has become intertwined with resource degradation.

Much of the ongoing loss of natural assets can be attributed to the lack of incentives and markets to provide compensation for the supply of essential environmental services, including hydrological services, carbon services, and biodiversity. There is, however, a growing recognition of the importance of these resources in the public domain.[6]

Environmental sustainability could become the next major challenge as India surges along its growth trajectory. Three striking messages emerge from this overview. First, most measured environmental indicators exhibit negative trends throughout the country and raise concerns about the effectiveness of the environmental policy regime and the efficiency of resource use. Especially noteworthy are declining water quality and increasing water scarcity, increases in cities with high and critical levels of air pollution (PM10), forest quality degradation and biodiversity loss, and land degradation. Second, environmental degradation has development impacts, suggesting that the management of environmental risks must be an important part of a growth strategy (see Box 1.1). Third, the unstoppable trends of urbanization, population growth, and industrialization mean that these environmental pressures are unlikely to abate for many decades.

Box 1.1 Development versus the Environment: Trade-Off or False Dichotomy?

Discussions of environmental problems in India mirror a longstanding debate on whether countries should focus on growth until poverty is eliminated. The position that they should is based on the view that environmental resources are a "luxury" that will be demanded (and affordable) as incomes rise with economic growth. It suggests that developing countries such as India should first accelerate economic growth and fix the environment at some time in the future. The "grow now, clean up later" doctrine, though much debated, is now widely discredited by the experiences of many developing countries.

At one extreme of this prolonged debate are reports that have emphasized the finiteness of resources and the limited capacity of the earth to absorb pollutants. The doomsday predictions reached particular prominence in the Club of Rome report *The Limits to Growth* (1972). The report looked at current trends of resource consumption and projected these into the future. It predicted that scarcity would lead to economic catastrophe by about 2020. However, there are no signs of economic collapse. The problem with the prediction is that the report neglected the important role of markets in assuring efficient resource allocation. When scarcity emerges, prices rise, consumption declines, and alternatives (substitutes) are found.

The next major global report was *Our Common Future* (1987) by the Brundtland Commission. It refrained from dramatic predictions but highlighted the degradation of the global commons, biodiversity, and other life-sustaining, nonmarketed assets. There are two main channels through which environmental factors could encumber growth. First is environmental quality—safe water and breathable air are among the benefits that development attempts to bring. If the benefits from growth are offset by these higher costs, the environmental impacts will retard development. Second, environmental damage can undermine productivity—soils, aquifers, forests, and ecosystems are all vital inputs needed to sustain economic activity. Resources that are depleted for current growth can jeopardize future economic prospects. The Brundtland report emphasized the role of market failure and the need for policies to address these issues in ways that are equitable and efficient. These are problems that still remain unresolved.

The most optimistic rebuttal to these reports came from an empirical relationship called the environmental Kuznets curve. Its proponents argued that global evidence showed that environmental degradation was just a matter of "growing pains" that would disappear with prosperity. This stylized fact is illustrated in Figure 1.1. The intuition of this statistical relationship may seem compelling. As an economy grows, the first issue to be addressed could be safe water—a major health hazard; next in line might be air pollution, a less visible but important health hazard; and somewhere in the future, biodiversity could be considered.

Whether this inverted U-shaped relationship between environmental quality and income levels actually holds in practice has been much debated in recent years. Most devastating for the proponents of the environmental Kuznets curve are findings from recent statistical analyses showing that the early estimates on which optimism was based were "spurious" (i.e., based on an invalid inference due to trending data). A significant finding that emerges from the literature is that the income-environment relationship is a matter of policy choice and is not predetermined. Good policies and effective institutions can deliver both higher incomes and a sustainable environment.

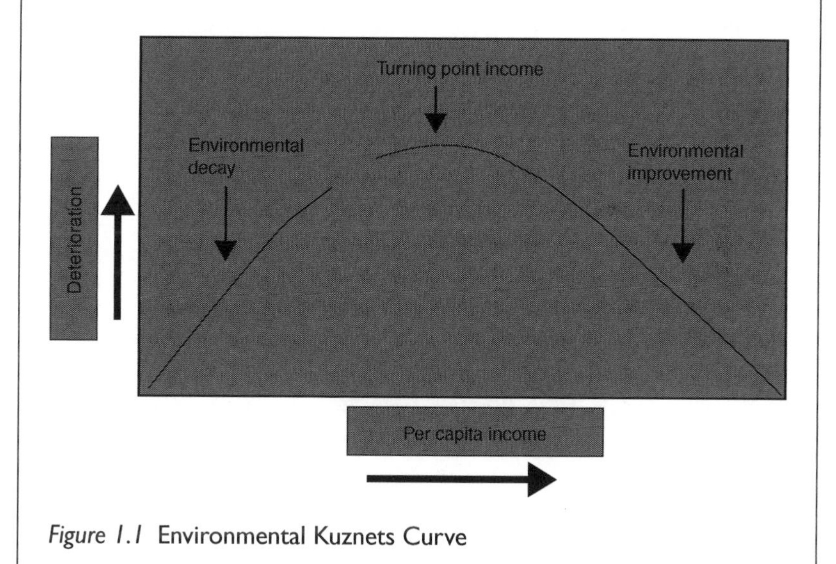

Figure 1.1 Environmental Kuznets Curve

Climate change poses an additional risk to long-term development prospects. It is projected that by the middle of the 21st century, the mean annual temperature in India will have increased 1.1–2.3°C under the moderate climate change scenarios (as per the A1B scenario of the IPCC).[7] All the global circulation models project that precipitation intensity[8] and heavy precipitation[9] events will increase, suggesting greater variability in rainfall.[10] The overall implication is that agro-climatic conditions will generally deteriorate across the country. The worst affected areas likely will be the arid and semiarid areas where agriculture is already under climate stress. Many of the major cities—Mumbai, Kolkata, Kochi, and Chennai—are located in low-lying coastal areas. Rising sea levels could disrupt economic activity, as well as the lives of some 100 million people living along the coastal belt. Further, striking impacts are likely to come from the melting Himalayan glaciers, which sustain agriculture and industry through the Gangetic plains.

1.2 Progress So Far

India has made substantial efforts in attempting to address environmental problems. The country has done many things right. It has enacted stringent environmental legislation and has created institutions to monitor and enforce legislation. There have been large-scale efforts to stabilize forest cover through afforestation schemes and considerable investments in water quality. India has enacted the National Environmental Policy (NEP), which recognizes the value of harnessing market forces and incentives as part of the regulatory approach.

Yet responses have not been commensurate with the scale of the challenge. The pace of change has undermined these investments in environmental protection. There are clear signs that keeping up with the environmental challenges of rapid urban growth, industrialization, and infrastructure development is proving to be a difficult task. Despite comprehensive environmental laws and severe penalties for violating pollution standards, offenders are seldom brought to justice. It is no surprise that in the absence of credible deterrents, enforcement of environmental laws remains weak. The relative outlay of government spending for the environment has also stagnated (see Figure 1.2). It has remained constant as a share of GDP and has been declining as a share of total public expenditure, even though pressures on the environment have been growing. The efficacy of environmental regulation has been further weakened by an overreliance on command and control policies that are conducive to rent-seeking behavior.

Although economic growth may create environmental pressures, the solution does not lie in slower growth. In India economic growth has helped create many of the preconditions for sound environmental management practices that have been adopted across

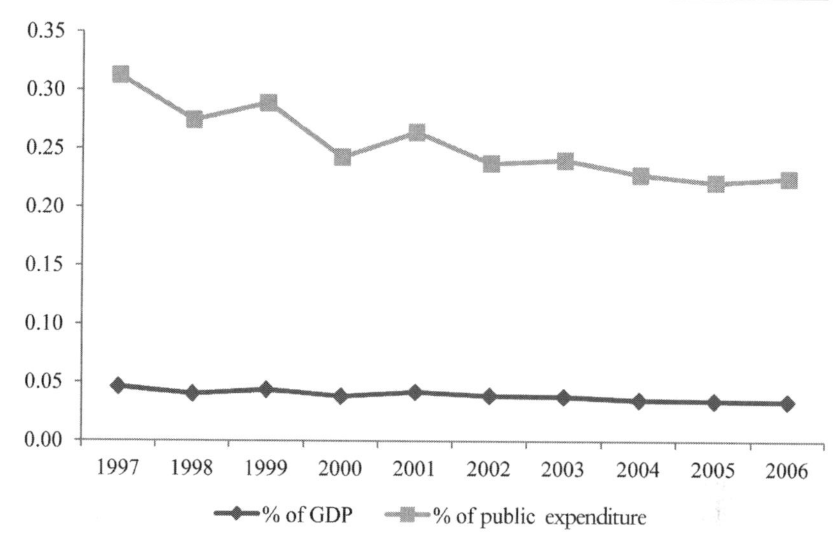

Figure 1.2 Plan outlay for environment as a share of GDP and total public expenditure

Source: World development indicators, World Bank (2010).

the country, and it has generated the policy space to address issues of sustainability. Policies for stronger growth often complement those for environmental protection—such as investments in clean water and sanitation or increases in the efficiency of resource use, particularly of common-property assets such as forests and shared water resources. Other problems are often exacerbated by economic expansion such as industrial pollution and the destruction of natural habitats. Here the challenge is to build incentives to address the environmental damage or externalities.

1.3 Study Objectives and Contribution

This study was commissioned by the government of India as a logical sequel to the India Country Environmental Analysis (CEA) report *Strengthening Institutions for Sustainable Growth*, published by the World Bank in 2007, which assessed the implementation effectiveness of environmental policies in specific key growth sectors: industry, power, and highways.

This study is cast in terms of the broader debate on the implications of rapid economic growth on environmental sustainability and the need to rethink our institutional arrangements in order to promote long-term environmentally sustainable development. The primary objectives of the study are as follows:

1. Provide estimates of social and financial costs of current environmental damage in India.
2. Identify conservation "hot spots" and provide an economic valuation of the ecosystem services across the country using quantified estimates of landscape types.
3. Assess trade-offs between economic growth and environmental sustainability for India using a computable general equilibrium (CGE) approach.

To achieve these objectives, the assessment proceeds at two levels: (1) a broad assessment of key environmental trends and their impacts on the economy and valuation of existing biodiversity and ecosystem services, and (2) an assessment of trade-offs between economic growth and environmental sustainability.

Following, therefore, are the specific study components:

1. A monetary valuation of environmental damage and quantification of environmental damage, using new techniques and methodologies.
2. Estimation of the value of ecosystem services from the major biomes in India using state-of-the-art methods with a view to preserving them for the future.
3. A menu of policy instruments, using a CGE approach, to explore trade-offs between economic growth and environmental sustainability with particular attention to air pollution.

The primary audiences for this report are the policy makers in the government of India, policy analysts, and think tanks. With its focus on the impacts of environmental degradation on economic growth, this study is one of the first of its kind in India and is aimed at quantifying and informing the government of India about the current and future environmental risks and constraints to its ambitious growth objectives. The report identifies the major environmental risks, assesses the likely economic costs, and suggests cost-effective and efficient policy responses to address these problems. The study therefore serves as a rigorous communication tool to help establish policy priorities and convey the message that environmental stewardship has development benefits.

The results should also be useful in guiding the World Bank's engagement by identifying environmental priorities, opportunities, and risks that need to be addressed in order to sustain high growth rates. It should also strengthen an entry point for establishing a dialogue on environmental issues with other agencies such as the Planning Commission as well as sectoral agencies and should inform the process leading to the implementation of the twelfth five-year plan.

Notes

1. *The Economist*, September 30, 2010.
2. Although the recent International Monetary Fund (IMF) survey of the Indian economy suggests a robust 7–8 percent growth in the next few years in spite of a global economic slowdown, it will be necessary, according to the IMF, to focus on reinvigorating the structural agenda, rather than relying on monetary and fiscal stimulus to ensure sustainable growth. Measures to facilitate infrastructure investment, reform the financial sector and labor markets, and address agricultural productivity and skills mismatches stand out (International Monetary Fund, 2012).
3. Data are from Indiastat.com.
4. Figures are from Indiastat.com; the definition of "middle class" is from the United Nations Environment Program (UNEP).
5. The annual study, the Environmental Performance Index, is conducted and written by environmental research centers at Yale and Columbia universities, with assistance from dozens of outside scientists.
6. The government of India appointed an expert group, headed by Sir Partha Dasgupta, in August 2011 to come up with a framework for greening the national accounts.
7. Based on the median values of an ensemble of 12 global circulation models (GCMs) with a standardized two-degree grid cell resolution; 2030–2049 compared to 1980–1999.
8. Precipitation intensity is the amount of rainfall per wet day, where a "wet" day is one with precipitation greater than 1 mm. The average increase across India is 4 percent (ensemble mean of 8 GCMs; 2030–2049 compared to 1980–1999).
9. As measured by the maximum amount of precipitation in any five-day period. The average increase across India is 11 percent (ensemble mean of 8 GCMs; 2030–2049 compared to 1980–1999).
10. The recent projections from the Indian Institute of Tropical Meteorology, one of the key government institutions studying climate change in India, in fact indicate that after 2050, temperatures will rise by 3–4 degrees over current levels, and rainfall will become both heavier and less regular, posing a grave threat to agriculture.
11. The just-released twelfth five-year plan calls for "faster, more inclusive, and *sustainable* growth" in its tagline.

Chapter 2

How Much Does It Cost?

2.1 Summary

That environmental degradation can be a by-product of economic activities is no secret. Industrial production often discharges pollutants into clean rivers and air, preventing the use of these resources for other purposes and harming the health of those exposed to them. Unsustainable agricultural practices can reduce crop productivity and cause dam sedimentation. Overexploitation of groundwater increases pumping costs and, if it leads to saltwater intrusion, may make aquifers unusable. These and many other forms of environmental degradation cause real costs to the economy and to people's welfare. Yet these costs often go unmeasured, and thus, their magnitude is largely unknown. Therefore, a country typically has insufficient information about the level of environmental damage, let alone information about the way to reduce or reverse the damage.

Until recently, most available studies have estimated the costs of environmental degradation for specific sites or industries. When government officials asked researchers a simple question about degradation—how large are the impacts of environmental degradation?—the response was often an emphatic "Large!" (a rather imprecise number). Since 2000, however, the World Bank has conducted a systematic effort to measure the cost of environmental degradation (COED) at the national and local levels in several countries (Croitoru and Sarraf, 2010). The strength of this type of work is that it actually quantifies in economic terms how large is "large" and thereby gains the attention of decision makers and offers specific insights for improved policy making.

This topical area provides estimates of social and financial costs of environmental damage in India from three pollution categories—(1) urban air pollution, including particulate matter and lead;

(2) inadequate water supply, poor sanitation, and hygiene; and (3) indoor air pollution—and three natural-resource damage categories—(1) agricultural damage from soil salinity, waterlogging, and soil erosion; (2) rangeland degradation; and (3) deforestation. The estimates are based on a combination of Indian data from secondary sources and on the transfer of unit costs of pollution from a range of national and international studies (a process known as benefit transfer). Data limitations have prevented estimation of degradation costs at the national level for coastal zones, municipal waste disposal, and inadequate industrial and hospital waste management. It is doubtful, however, that costs of degradation and health risks arising from these categories are anywhere close to the costs associated with the categories considered. Furthermore, the estimates provided do not account for loss of non-use values (i.e., values natural resources hold for people even when they do not use them). These could be important, but there is considerable uncertainty about the values.

Methodology for Valuation of Environmental Damage

The quantification and monetary valuation of environmental damage involves many scientific disciplines, including the environmental, physical, biological, and health sciences; epidemiology; and environmental economics, which relies heavily on other fields within economics, such as econometrics, welfare economics, public economics, and project economics. New techniques and methodologies have been developed in recent decades to better understand and quantify preferences and values of individuals and communities in the context of environmental quality, conservation of natural resources, and environmental health risks. The results from these techniques and methodologies can then be, and often are, utilized by policy makers and stakeholders in the process of setting environmental objectives and priorities. And because preferences and values are expressed in monetary terms, the results provide some guidance for the allocation of public and private resources across diverse sectors in the course of socioeconomic development.

The terminology used here needs some qualification. "Environmental damage" means physical damages that have an origin in the physical environment. Thus, damages to health from air or water pollution are included, as well as damages from deforestation. The term "cost" means the opportunity cost to society—that is, what is given up or lost when a course of action is taken. When goods traded in markets are damaged, prices and knowledge of consumer preferences for the damaged goods (embodied in the demand function), as well as production information (embodied in the supply function), provide the necessary information for computing social costs. Estimating social costs from reduced productivity of agricultural land due to erosion, salinity, or other forms of land degradation is a good example. However, many damages from environmental causes are to "goods" that are not traded in markets, such as health. In these cases, economists have devised a number of methods for estimating social costs based on derived preferences from observable or hypothetical behavior and choices.

One example is the value of time lost to illness or provision of care for ill family members. If the person who is ill or who is providing care for someone who is ill does not otherwise have a job, the financial cost of time lost is zero. However, even in such a case, the person normally would be engaged in activities that are valuable for the family, and illness reduces the amount of time available for these activities. Thus, there is a social cost of time losses to the family. In an economic costing exercise, this is normally valued at the opportunity cost of time—that is, the salary or fraction of the salary that the individual could earn if he or she chose to work for income. In summary, social costs are preferred over financial costs because social costs capture the cost and reduced welfare to society as a whole. All costs are estimated as flow values (annual losses).

Unfortunately, the information needed to estimate social costs for some categories is often lacking, particularly in developing countries such as India. In such cases one has the option of relying on financial costs, which generally do not capture all the social costs. In this report, financial costs have been used for a significant part of the analysis, but social costs are reported wherever these could be obtained or estimated. In general for a country such as India, these financial costs are likely to underestimate social costs.

Interpretation of Results

The methodology of COED estimations is close to the green accounting concept, yet it is not the same. Whereas green accounting takes into account positive and negative changes, COED focuses on a negative side only. This methodology is widely used in the World Bank and aims to communicate the current level of the negative impact on environment and natural resources. There is an ongoing effort to create an inclusive system of green accounting for India (Dasgupta, 2011) that is methodologically different from this study.

Estimates of the costs of degradation are generally reported as a percentage of conventional GDP. This provides a useful estimate of the importance of environmental damages, but it should not be interpreted as saying that GDP would increase by a given percentage if the degradation were reduced to zero. Any measures to reduce environmental degradation would have a cost, and the greater the reduction made, the higher the cost. Hence, a program to remove all degradation could well result in a lower GDP. The analysis of the "right" level of reduction is an additional exercise that is not part of this (or indeed any) cost of degradation study. What is provided here is a measure of the overall damage relative to a benchmark, in which all damages related to economic activity are eliminated.

The benchmark clearly has a major effect on the estimates produced. The aim in each case is to assess the level of damage that can be attributed to economic activity, but this is not always easy to establish, and there is always an element of arbitrariness in the value chosen. In this report we give the benchmark value of each category of damages, with whatever justification is available. We also try to be consistent with benchmark values used in similar studies for other countries that have been conducted at the World Bank. Table 2.1 summarizes the benchmark values used in the study.

Uncertainty

The exercise conducted here involves a great deal of uncertainty, including that arising from limitations of data on social costs, from methods used to estimate the effects of pollution and resource degradation on indicators of health or output (i.e., the concentration-response functions), and from the transfer of some unit values from studies outside of India. It would be a major task to handle all these uncertainties quantitatively, and that has not been possible in this study. In particular, to keep the analysis simple, we do not report all the statistical uncertainties, such as those

Table 2.1 Benchmark values used in the study

Source of damage	*Benchmark value*	*Comment*
Health		
Mortality from PM2.5	7.5 ug/m3	Assumed background level in many studies including WHO
Morbidity from PM10	Zero concentration	
Exposure to lead	10 ug/dl	WHO methodology (Pruss-Ustin et al., 2004)
Mortality and morbidity from waterborne diseases	Disease rates that prevail in developed countries	WHO methodology (Fewtrell and Colford, 2004)
Averting expenditures against unsafe water	Zero	No expenditure is necessary if water supply is safe
Mortality and morbidity from indoor air pollution	Odds Ratio of 1	Implies no additional risk of these impacts as a result of indoor air pollution
Natural resources other than forests		
Soil salinity and waterlogging	Zero salinity/ waterlogging	No loss of productivity compared to unaffected areas
Soil erosion	Zero erosion	No soil loss
Rangeland	Zero loss	No loss of productivity compared to unaffected areas
Forest degradation		
Timber	Value of service in nondegraded forest	80–100% loss
Nontimber products		20–100% loss
Ecotourism		100% loss

Source: Staff estimates.

for concentration-response coefficients, and we rely on central estimates. Although some components of the central estimates do use "mean" input parameters and estimates, some inputs into the damage calculations cannot be considered "means" in the statistical sense. For example, they may be judgmental estimates based on a mixture of the expected mean or median values. Thus, the reader should interpret these estimates as "midpoint" or "middle" values. At the same time we have attempted to represent the uncertainty for each category of damage by providing a range based on a combination of factors, details of which can be found in the relevant sections.

Finally, in making the estimates, we have taken a conservative approach or, put another way, a "defensible borders" approach, where we chose models and data and make assumptions and interpretations that, at least partly, are justified by pointing out that other approaches would yield higher estimates of social costs.

Results

The report estimates the total cost of environmental degradation in India at about Rs 3.75 trillion (US$80 billion) annually, equivalent to 5.7 percent of GDP in 2009, which is the reference year for most of the damage estimates. High and low estimates for the selected degraded media are presented in Table 2.2.

Table 2.2 Annual cost of environmental damage—low and high estimates (Rs billion per year)

	"Low"	*Midpoint estimate*	*"High"*	*Midpoint estimate as percent of total cost of environmental damage*
Environmental Categories				
Outdoor air pollution	170	1,100	2,080	29%
Indoor air pollution	305	870	1,425	23%
Cropland degradation	480	703	910	19%
Water supply, sanitation, and hygiene	475	540	610	14%
Pasture degradation	210	405	600	11%
Forest degradation	70	133	196	4%
Total annual cost	1,710	3,751	5,821	
Total as percent of GDP in 2009	2.60%	5.70%	8.80%	

Note: Staff estimates are rounded to the nearest 10.

2.2 Cost of Environmental Degradation

Environmental pollution, degradation of natural resources, natural disasters, and inadequate environmental services, such as poor water supply and sanitation, impose costs to society in the form of ill health, lost income, and increased poverty and vulnerability. This section provides overall estimates of the social and financial costs of such damages, referring as much as possible to damages for 2009. In some cases, however, the figures may be based on damages in an earlier year if that was the latest information available.

Of all the categories of degradation listed previously, only natural disasters are not the result of anthropogenic factors, although some argue that human impacts on the environment are causing an increase in the incidence and severity of so-called natural disasters. We do not include them in the main set of estimates because of these varying views. Since the damages arising from natural disasters are of interest to policy makers, and some COED studies do include them, we have reported these damages separately in Appendix 3.

The results are summarized in Figure 2.1 and Figure 2.2 and in Table 2.2. Total damages amount to about Rs 3.75 trillion (US$80 billion), equivalent to 5.7 percent of GDP. Of this total, outdoor

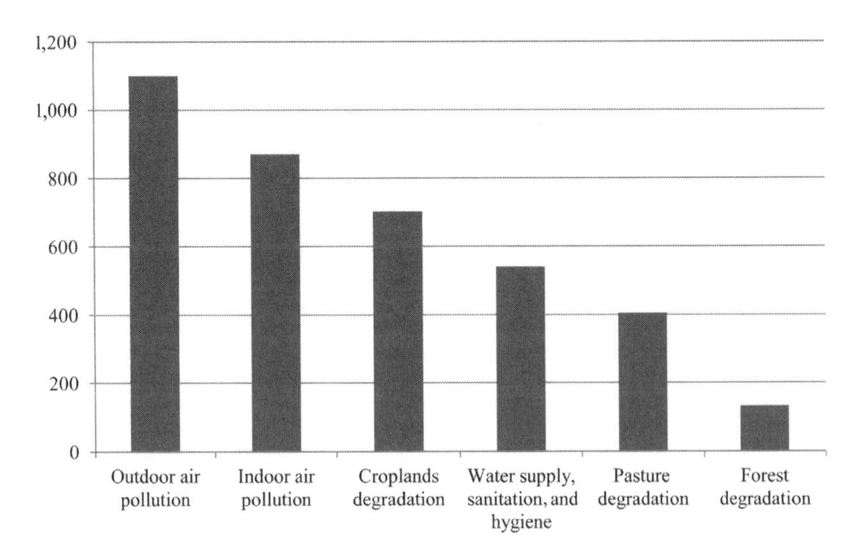

Figure 2.1 Annual cost of environmental damage (Rs billion)

Source: Staff estimates.

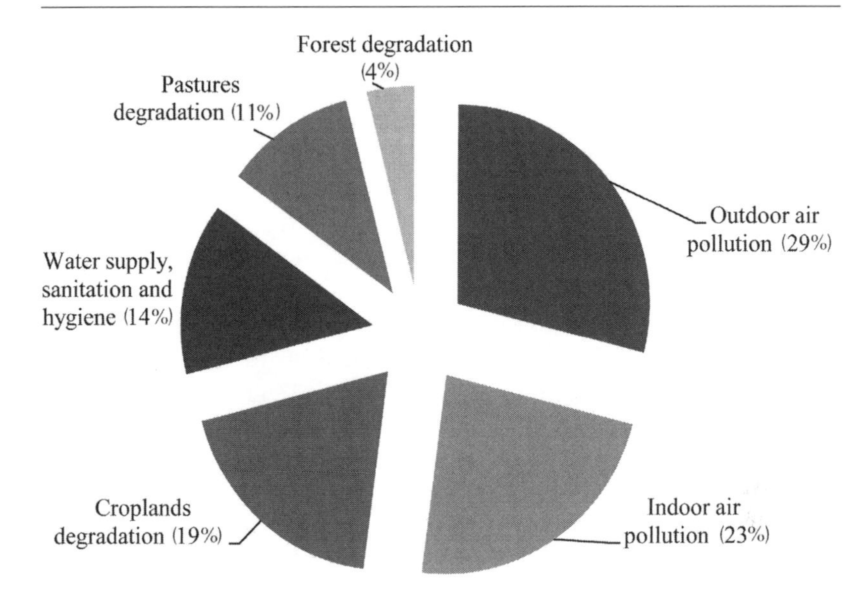

Figure 2.2 Relative share of damage cost by environmental category

Source: Staff estimates.

air pollution accounts for the highest share at Rs 1.1 trillion (Figure 2.1), followed by the indoor air pollution cost at Rs 0.9 trillion; the croplands degradation cost at just over Rs 0.7 trillion; the inadequate water supply, sanitation, and hygiene cost at around Rs 0.5 trillion; the pasture degradation cost at Rs 0.4 trillion; and the forest degradation cost at Rs 0.1 trillion. The individual damages are shown as shares of the total in Figure 2.2. Outdoor air pollution accounts for 29 percent of the total damages, followed by indoor air pollution (23 percent), cropland degradation (19 percent), inadequate water supply and sanitation (14 percent), pasture degradation (11 percent), and forest degradation (about 4 percent).

In addition India has experienced some damages from natural disasters (floods, landslides, tropical cyclones, and storms). These are not included in the preceding figures for the reasons previously given. Over the period 1953–2009 damages from natural disasters were estimated at Rs 150 billion a year on average (in constant 2009 prices) and took the form of injuries and loss of life, losses to livestock and crops, and losses to property and infrastructure. Details are given in Appendix 2.[1]

In addition to the midpoint values, low and high estimates of annual costs are presented in Table 2.2. The low and high range estimates differ considerably across the categories because of the uncertainties related to economic valuation procedure or uncertainties about exposure to specific hazards. The urban air pollution estimate range is mainly affected by the social cost of mortality, which is derived by applying two different valuation techniques (section 2.3). The range for indoor air pollution arises mainly from the uncertainty about exposure level to indoor smoke and from the use of fuel wood (section 2.5). In the case of agricultural soil degradation, the range is associated with uncertainty about yield losses from salinity (section 2.6). The range for water supply, sanitation, and hygiene is in large part associated with uncertainties regarding estimates of child mortality and morbidity from diarrhea (section 2.4). The range for deforestation is associated with the uncertainty of the use benefits of forest (section 2.7). If we take the lower bound of the estimates, the figures are about 45 percent of the mean values (or 2.6 percent of GDP), whereas if we take the upper bound, they are 64 percent higher than the mean (or about 8.8 percent of GDP).[2]

Health–Related Damages among Selected Populations in India

The damages associated with environmental health are estimated for different groups of the population. The outdoor air pollution losses were estimated for the inhabitants of cities with a population of more than 100,000 (due to data limitations); inadequate water supply, sanitation, and hygiene costs were estimated for the whole population of India; and indoor air pollution costs were estimated for the households that use solid fuel for cooking (about 75 percent of all households). These differences in coverage should be borne in mind when comparing across the different environmental burdens. In particular, coverage for outdoor air pollution is less complete than the others, and thus the figures for that category are underestimated.

The higher costs for outdoor and indoor air pollution are primarily driven by elevated exposure of the urban and rural populations to particulate-matter pollution that results in a substantial cardiopulmonary and chronic obstructive pulmonary disease mortality load among adults. As noted, the rural population has been assessed only for indoor air pollution.

Figure 2.3 gives estimates of damage per person within the different exposed populations used to construct the figures in Table 2.2.

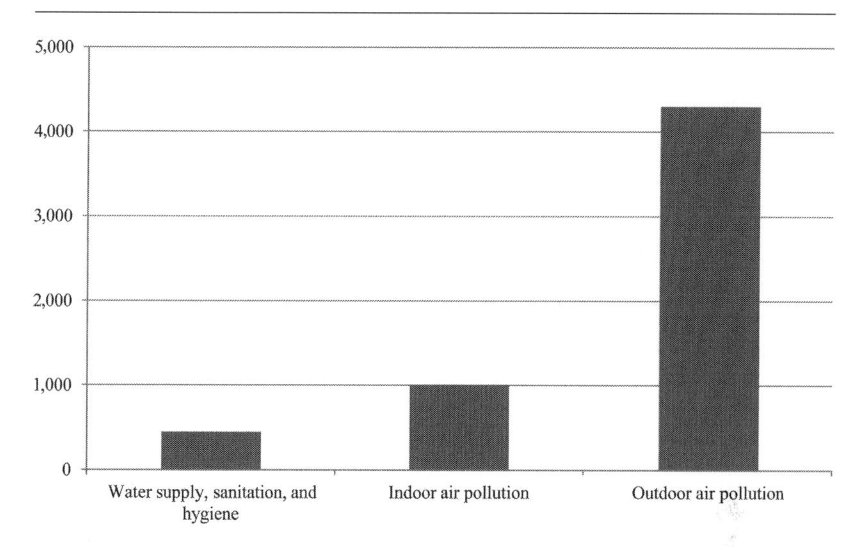

Figure 2.3 Annual environmental health losses per person of the exposed population (Rs.)

Source: Staff estimates.

We note that a significant part of the health burden, especially from water supply, sanitation, and hygiene, is borne by children under age five years (Figure 2.4). These figures suggest that about 23 percent of under-five mortality can be associated with indoor air pollution and inadequate water supply, sanitation, and hygiene, and 2 percent of adult mortality is associated with outdoor air pollution.

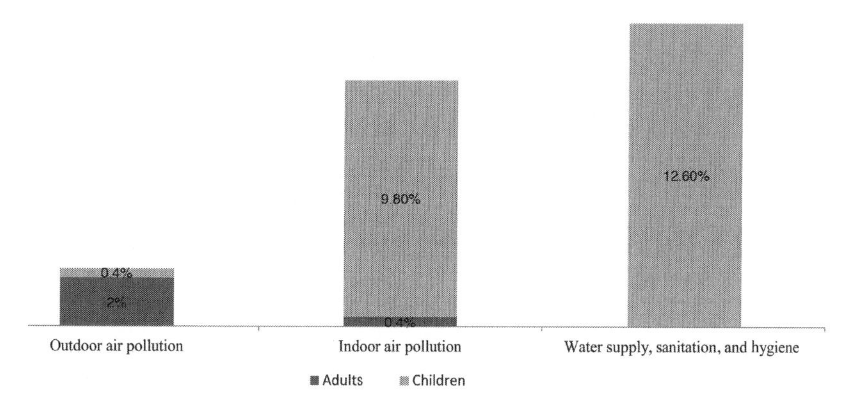

Figure 2.4 Estimated share of annual mortality from different sources in India

Source: Staff estimates.

Environmental Damages and the Poor

Although this study does not address the impacts of the previously estimated losses on poor households specifically (that is something that should be undertaken as a separate study), one can comment on how the poor are affected by the environmental damages. First, the losses related to water, sanitation, and hygiene are likely to be concentrated among the poor, who most often do not have access to piped water or sanitation. Second, the rural population is more affected by water and indoor air pollution–related damages than the urban population. For the urban population the distribution of impacts by income class is less certain. Some studies indicate that urban ambient air quality does affect the poor more than the rich (Garg, 2011), but the present study has not been able to confirm this point. In overall terms, however, it is very likely that the poorer urban population suffers more both from urban air pollution and from inadequate water supply, sanitation, and hygiene, and in general the poor are included in all major cost categories (those who live in big cities and use solid fuel for cooking).

Other Categories of Damages

Cropland damages arise from the decline in the value of agricultural outputs resulting from yield losses due to soil erosion, waterlogging, salinity, and overgrazing. We derive a range of estimates because of the uncertainty of crop and pasture profitability as well as the uncertainty of the level of degradation.

Forest degradation has arisen in India from unsustainable logging practices in some regions and from general overexploitation of forest resources. Although the country gained about 7 percent in overall forest cover between 1990 and 2010, there has also been a notable degradation in some forests. This results in losses of ecosystem services, including carbon sequestration, provision of timber and nontimber forest products, recreational and cultural use of forests, and prevention of soil erosion. The losses are valued using a range of techniques, which are subject to considerable uncertainty arising from the estimates of forest productivity and the methods of obtaining values for the nonmarketed services.

Finally, impacts of changes in fisheries were examined, but it was not possible to value these impacts in monetary terms because of gaps in the data.

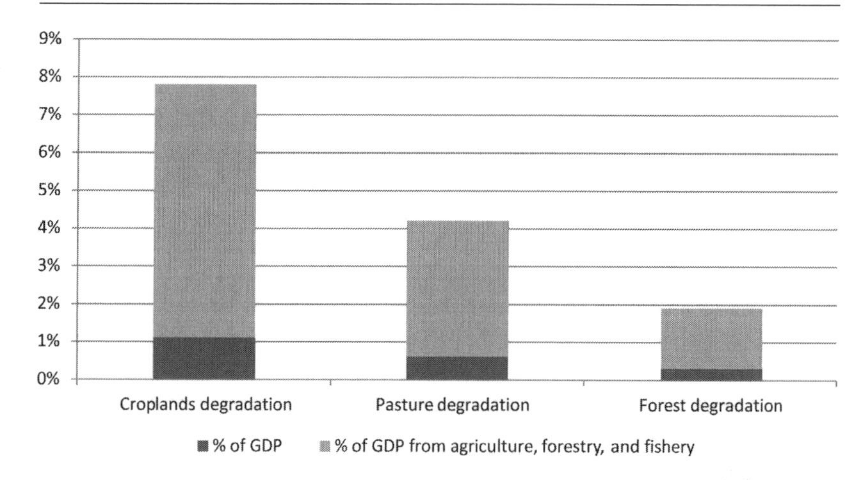

Figure 2.5 Natural resource losses compared to GDP and GDP in agriculture, forestry, and fishery in 2009

Source: Staff estimates.

Another way of looking at the role of environmental resources is in terms of the "GDP of the poor."[3] Natural resources degradation is more significant when compared with the income of the poor. One measure of the growth potential for the poor is in the share of GDP generated in agriculture, forestry, and fishery, which made up about 17 percent of GDP in 2010. Figure 2.5 summarizes potential impacts of natural resource degradation losses on the overall GDP and the "GDP of the poor" (i.e., GDP in agriculture, forestry, and fishery). In total these losses amount to about 2 percent of total GDP but 11 percent of the "GDP of the poor" in India. It should be noted that although estimating impacts on the "GDP of the poor" is an interesting concept, this could also be an underestimation of the impact of environmental damage suffered by the poor given that much of the health damage from pollution in urban areas is also predominantly borne by the urban poor.

Comparison with Other Countries

The cost of environmental degradation in India is roughly comparable to other countries with similar income levels (Figure 2.6). Studies of the cost of environmental degradation were conducted using a similar methodology in Pakistan, a low-income country, and in several low- and lower-middle-income countries in Asia, Africa, and Latin America. These studies show that the monetary

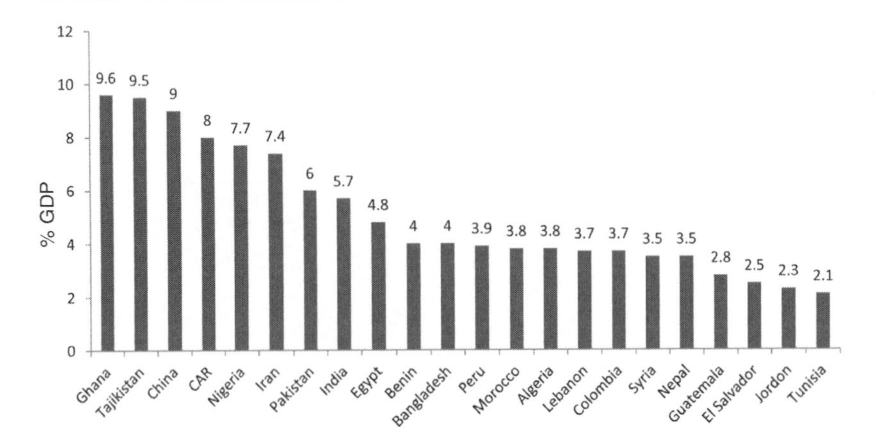

Figure 2.6 Cost of environmental degradation (Health and natural resources damages)

Source: World Bank (2012), *Inclusive Green Growth:The Pathway to Sustainable Development.*

value of increased morbidity, mortality, and natural resources degradation typically amounts to 4–10 percent of the GDP, compared to 5.7 percent of GDP in India.[4]

The situation also looks consistent across different countries if one compares only the health costs of outdoor air pollution (Figure 2.7).

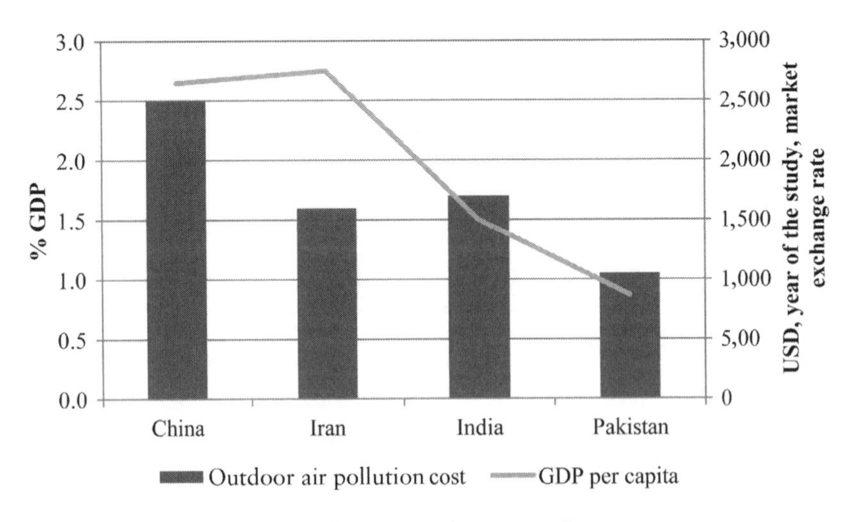

Figure 2.7 Health cost attributed to outdoor air pollution

Sources: Washington, D.C.: World Bank (2005), *Islamic Republic of Iran: Cost Assessment of Environmental Degradation*; World Bank (2006), *Ghana Country Environmental Analysis*; World Bank (2007), *Pakistan: Country Strategic Environmental Assessment*; World Bank (2007), *Cost of Pollution in China: Economic Estimates of Physical Damages.*

In all the selected countries, these costs vary between 1.1 and 2.5 percent of GDP. In India the health cost of outdoor air pollution is estimated at about 1.7 percent of GDP. The high cost of outdoor air pollution–related mortality in urban areas is the main driver of environmental health costs.

A World Bank (2007) study on China, later cited in *China 2030* (World Bank, 2012), applied a methodology for outdoor air pollution valuation similar to the one used in this report.

2.3 Urban Air Pollution

Particulate Matter

There is substantial research evidence from around the world showing that outdoor urban air pollution has significant negative impacts on public health, including premature deaths, chronic bronchitis, and respiratory disorders. A comprehensive review of such studies is provided in Ostro (1994, 2004). The air pollutants that have shown the strongest association with these health results are particulate matter and other secondary particles with similar characteristics of less than 10 microns in diameter (PM10).[5] Research in the United States in the 1990s and most recently by Pope and colleagues (2002) provides strong evidence that particulates of less than 2.5 microns in diameter (PM2.5) have the largest health effects. Other gaseous pollutants (SO_2, NO_x, CO, and ozone) are generally not thought to be as damaging as fine particulates. However, SO_2 and NO_x may have important health consequences because they can react with other substances in the atmosphere to form secondary particulates. The evidence implicates sulfates formed from SO_2 but is much less certain about nitrates formed from NO_x.

The focus here, therefore, is the health effects of all fine particulates (PM10 and PM2.5).[6] This requires data on who is exposed, the health impacts of that exposure, and the value attached to those impacts.

Given data limitations, we can only estimate impacts for the urban populations and in fact only for a part of that population. Only major cities have total suspended particulates (TSP) and PM10 monitoring data. In this study we focus only on cities with a population of at least 100,000. Since the baseline population is from the 2001 census, there are many cities that have achieved a population of 100,000 in the meantime but have not been included in the study. This can be updated in the future.

Pollution data for all cities, where available, were taken from the Central Pollution Control Board's (CPCB) Environmental Data Bank website for the year 2008. Health damage estimates for PM10 were calculated based on observations for the year 2008. The study included 96 cities with monitoring stations and about 225 cities with no monitoring stations (254 million people in total). The population for the 96 cities with monitoring stations amounts to 186 million, or about 16 percent of the country's population. The estimated annual health effects on the population in these cities are given in Appendix 1 (Table A1.1), which provides details on estimates of exposed urban population and annual average PM10 levels. In addition there are about about 225 cities with—a total population of 69 million for which there are no data on PM concentrations. Since excluding them from the estimation of health impacts would be a serious omission, annual average PM10 levels were assigned to these cities based on scaling up of the World Bank model for PM10 concentrations (taken from the World Bank internal research database). Appendix 1 (Table A1.3) lists the additional cities included and the estimated concentrations.

The age distribution of the urban population was estimated using urban population parameters from the 2001 India Census. PM10 values were transformed into PM2.5 values using a ratio of 0.5 based on evidence from India (Central Pollution Control Board, 2011). This ratio reflects the mean of the PM2.5/PM10 ratio for large Indian cities reported in this chapter.

Based on the current status of worldwide research, the risk ratios—or concentration-response coefficients from Pope and colleagues (2002)—were considered likely to be the best available evidence of the mortality effects of ambient particulate pollution (PM2.5).

Damages caused by anthropogenic factors are measured from a baseline PM2.5 concentration, which we set equal to 7.5 $\mu g/m^3$ (as in World Health Organization, 2002a). This is considered to be the level one would find in the natural environment. A log-linear function for estimating cardiopulmonary mortality associated with outdoor air pollution was applied. The methodology is described in Appendix 1.

The morbidity effects assessed in most worldwide studies are based on PM10. Concentration-response coefficients from Ostro and Chestnut (1998) and Abbey et al. (1995) have been applied to estimate these effects. Ostro reviews worldwide studies (1994), and based on that, estimates concentration-response coefficient for restricted activity days (1998), and Abbey et al. (1995) provides estimates of chronic bronchitis associated with particulates (PM10). A linear function for estimating morbidity endpoints associated with outdoor air pollution was applied. The methodology is described in Appendix 1.

The mortality and morbidity coefficients are presented in Table 2.3 based on these estimates. Further details on the application of the concentration-response coefficients are given in Appendix 1.

The health effects of air pollution can be converted to disability-adjusted life years (DALYs) to facilitate a comparison with health effects from other environmental risk factors. DALYs per 10,000

Table 2.3 Urban air pollution concentration-response coefficients

Annual health effect	*Concentration-response coefficient*	*Per 1 μg/m³ annual average ambient concentration of*
Long-term mortality (% change in cardiopulmonary and lung cancer mortality)	0.8%*	PM2.5
Acute mortality children under five (% change in ARI deaths)	0.166%	PM10
Chronic bronchitis (% change in annual incidence)	0.9%	PM10
Respiratory hospital admissions (per 100,000 population)	1.2	PM10
Emergency room visits (per 100,000 population)	24	PM10
Restricted activity days (% change in annual incidence)	0.475%	PM10
Lower respiratory illness in children (per 100,000 children)	169	PM10
Respiratory symptoms (per 100,000 adults)	18,300	PM10

* Mid-range coefficient from Pope et al. (2002) reflecting a linear function of relative risk. In the analysis however, we used a log-linear.

Sources: Pope et al. (2002) and Ostro (2004) for the mortality coefficients; Ostro (1994, 1998) and Abbey et al. (1995) for the morbidity coefficients.

Table 2.4 DALYs for different health endpoints

Health effect	*DALYs lost per 10,000 cases*
Mortality: adults	75,000
Mortality: children under 5	340,000
Chronic bronchitis (adults)	22,000
Respiratory hospital admissions	160
Emergency room visits	45
Restricted activity days (adults)	3
Lower respiratory illness (children)	65
Respiratory symptoms (adults)	0.75

cases of various health endpoints are presented in Table 2.4. Further details on how they were arrived at are given in Appendix 1.

Urban air particulate pollution is estimated to cause around 109,000 premature deaths among adults and 7,500 deaths among children under five years annually. This estimated adult mortality is consistent with Cropper et al.'s (2012) estimate of the annual mortality associated with coal electricity generation in India (about 60,000 people, calculated as about 650 deaths per year for each of the 92 coal-burning power plants in India). Electricity generation is responsible for a fraction of PM pollution analyzed in this report.[7] Estimated new cases of chronic bronchitis are about 48,000 per year. Annual hospitalizations due to pollution are estimated at close to 370,000, and emergency room visits/outpatient hospitalizations are estimated at 7.3 million per year. Cases of less severe health impacts are also presented in Table 2.5. In terms of annual DALYs, lost mortality accounts for an estimated 60 percent, chronic bronchitis for around 5 percent, restricted activity days for 7 percent, and respiratory symptoms for 25 percent.

The estimated annual cost of urban air pollution health effects is presented in Table 2.6. The cost of mortality is based on the human capital approach (HCA) as a lower bound and the value of statistical life as an upper bound for adults and on HCA for children. Details of the valuation of mortality and morbidity endpoints are given in Appendix 1.

The cost-of-illness approach (mainly medical costs and value of time losses) was applied to obtain an estimate of the morbidity cost (see cost of morbidity in Table 2.6).

To summarize, the mean estimated annual cost of PM urban air pollution totals Rs 1.103 billion or 1.7 percent of GDP in 2009.

Table 2.5 Estimated health impact of urban air pollution

Health endpoints	*Total cases*	*Total DALYs*
Premature mortality (adults)	109,340	820,049
Mortality (children under 5)	7,513	255,431
Chronic bronchitis	48,483	106,663
Hospital admissions	372,331	5,957
Emergency room visits/Outpatient hospital visits	7,303,897	32,868
Restricted activity days	1,231,020,030	369,306
Lower respiratory illness in children	16,255,360	105,660
Respiratory symptoms	3,917,855,052	293,839
Total		1,989,773

Table 2.6 Estimated annual cost of health impacts (Rs billion)

Health categories	*Total annual cost (billion Rs)*	*Percent of total cost*
Mortality:		
Adults	1,018	92.2%
Children under age 5	13	1.2%
Morbidity:		
Chronic bronchitis	1	0.1%
Hospital admissions	3	0.3%
Emergency room visits/ outpatient hospital visits	8	0.7%
Restricted activity days (adults)	46	4.2%
Lower respiratory illness (children)	14	1.3%
Total cost of morbidity	72	6.6%
Total cost (mortality and morbidity)	1,103	100%

About 93 percent of the cost is associated with mortality, and 7 percent with morbidity. Measured in terms of DALYs,[8] about 54 percent of the cost is associated with mortality and 46 percent with morbidity (Table 2.3). All damages are measured from a baseline concentration of PM2.5 of 7.5 μg/m^3 and zero threshold of PM10. More details on the methodology of the analysis are presented in Appendix 1.

2.4 Water Supply, Sanitation, and Hygiene

The main health impacts of unclean water and poor hygiene are diarrheal diseases, typhoid, and paratyphoid. In addition there are costs in the form of averting expenditures to reduce health risk. Diarrheal and related illnesses contribute the dominate share of the health cost.

Diarrheal Diseases, Typhoid, and Paratyphoid

Based on an extended meta-analysis of peer-reviewed publications, the World Health Organization (WHO) has proposed a rigorous methodology[9] that links access to improved water supply, safe sanitation, and hygiene to incidence of diarrheal mortality and morbidity of children under five years old and other population morbidity. About 88 percent of diarrheal cases globally are attributed to water, sanitation, and hygiene (Prüss-Üstün, Fewtrell, and Landrigan, 2004). This is a conservative approach where malnutrition impact on early childhood diseases is omitted. If considered, this additional indirect impact would approximately double the mortality attributed to water supply, sanitation, and hygiene (World Bank, 2010). However, a number of these losses are in the form of acute respiratory mortality, which was accounted for in the indoor air pollution section 2.5. To avoid double counting and to be on the conservative side, in this section we consider only direct impact of inadequate water supply, sanitation, and hygiene.

Mortality for children under five and diarrheal-induced child mortality are high in India. Baseline health data for estimating the health impacts of inadequate water supply, sanitation, and hygiene are presented in Table 2.7. The Office of the Registrar General indicated in 2004 that 14 percent of child mortality was due to intestinal diseases. A baseline diarrheal mortality rate of 14 percent for under-five child mortality is thus used for diarrheal mortality estimation.

For diarrheal morbidity, however, it is very difficult or practically impossible to identify all cases of diarrhea. The main reason is that substantial numbers of cases are not treated or do not require treatment at health facilities and are therefore never recorded. A second reason is that cases treated by private doctors or clinics are often not reported to public health authorities. Household surveys therefore provide the most reliable indicator of total cases of diarrheal illness. Most household surveys, however, contain information on diarrheal illness only in children. Moreover, the surveys reflect diarrheal prevalence only at the time of the survey. Because there is often high variation in diarrheal prevalence across seasons of the year, extrapolation to an annual average will result in either an overestimate or underestimate of total annual cases. Correcting this bias is often difficult without knowledge of seasonal variations.

In spite of all these difficulties, a reasonable estimate has been made of the number of cases and prevalence of diarrhea in the population, along with the number of DALYs per 100,000 cases of diarrhea. Details are given in Appendix 1, with the figures summarized in Table 2.7.

Table 2.8 presents the estimated health impacts from inadequate water, sanitation, and hygiene, based on the parameters given in Table 2.7, including the assumption (from WHO) that 88 percent of diarrheal illness is attributable to water, sanitation, and hygiene. The table also provides estimates of DALYs lost to waterborne diseases. About 60 percent of the DALYs are from diarrheal child mortality. Typhoid/paratyphoid deaths add another 20 percent of DALYs.

The estimated costs associated with the impacts identified here are given in Table 2.9. Details of the baseline cost data are given in Appendix 1 (Table A1.6). The hypothetical values on which the estimates are based rely on the WHO methodology, which uses conditions in developed countries as the benchmark. The incidence rates for these illnesses are close to zero in those countries (0.3 per person per year, as in Fewtrell and Colford, 2004). Further details are given in Appendix 1.

The total cost is Rs 490 billion. The cost of mortality is based on the human capital approach (HCA) for children under five (see Appendix 1.5). The cost of morbidity includes the cost of illness (medical treatment, medicines, and value of lost time) and value of lost DALYs estimated at GDP per capita. We used GDP per capita as a proxy for willingness to pay (WTP) for one additional year of life, expressed in DALYs.

Averting Expenditures

In the presence of perceived health risks, individuals often take measures to avoid these risks. These measures are usually considered a cost of the health risks of environmental burdens. If consumers perceive that the municipal water supply or the other sources

Table 2.7 Baseline data for estimating health impacts

	Baseline	*Source*
Child mortality rate for those younger than age 5 years in 2006	52–82 (per 1,000 live births)	NFHS-3
Diarrheal mortality in children younger than 5 years (% of child mortality)	14 %	Office of Registrar General (2004)
Diarrheal two-week prevalence in children younger than 5 years	8.9–9%	NFHS-3
Estimated annual diarrheal cases per child younger than 5 years	1.85–1.87	Estimated from NFHS-3
Estimated annual diarrheal cases per persons older than 5 years	0.37–0.56	International experience (Krupnick et al., 2006)
Hospitalization rate (% of all diarrheal cases) for children younger than 5 years	0.15%	NSS (2004)
Hospitalization rate (% of all diarrheal cases) for the population older than 5 years	0.3–0.6%	
Percent of diarrheal cases attributable to inadequate water supply, sanitation, and hygiene	90%	WHO (2002b)
DALYs per 100,000 cases of diarrhea in children younger than 5 years	70	Estimated from WHO tables
DALYs per 100,000 cases of diarrhea in the populations older than 5 years	100–130	
DALYs per 100,000 cases of typhoid in the entire population	190–820	
DALYs per case of diarrheal and typhoid mortality in the entire population	32–34	

Table 2.8 Estimated annual health impacts from Water, Sanitation, Hygiene

	Cases		*Estimated annual DALYs (thousands)*		*Percent of total DALYs*
	Urban	Rural	Urban	Rural	
Children under 5 years: increased mortality	41,000	198,000	1,384	6,714	93%
Children under the age of 5 years: increased morbidity	57,831,000	178,898,000	20	63	1%
Population over 5 years of age: increased morbidity	149,836,000	344,183,000	177	406	6%

Source: Staff estimates.

Table 2.9 Estimated health impacts from inadequate Water, Sanitation, Hygiene

	Estimated annual cost (billions Rs)		
	Urban	*Rural*	*Total*
Mortality			
Children younger than age 5 years: diarrheal mortality	50	227	277
Children younger than age 5 years: typhoid			0.3
Persons older than age 5 years: typhoid			0.5
Morbidity			
Diarrheal (all ages)	105	103	208
Typhoid*			3.3
Total annual cost	155	330	489.1

* About 25 percent of estimated COI is from hospitalization and doctor visits, and 70 percent is from time losses for the ill individuals and their caregivers during illness.

Source: Staff estimates.

Table 2.10 Estimated total annual household cost of averting expenditures

	Total annual cost (billion Rs)	
	Urban	*Rural*
Cost of bottled water consumption	20	7
Cost of household boiling drinking water	4	3
Cost of household filtering drinking water	14	7
Total annual cost	38	17

Source: Staff estimates.

of water supply they rely on are unsafe, they are likely to purchase bottled water for drinking purposes, boil their water, or install water purification filters. The estimated costs of these options are given in Table 2.10, with details on the estimated unit costs available in Appendix 1. The assumed hypothetical level of expenditure here is zero (i.e., no aversive expenses would be incurred if the water supplied was safe). The total aversive expenditures for India amount to about Rs 55 billion per year.

In summary the estimated annual cost associated with inadequate water supply, sanitation, and hygiene is presented in Figure 2.8, totaling Rs 470–610 billion per year, with a mean of Rs 540 billion. The cost of health impacts represents an estimated 90 percent of total mean cost, with averting expenditures accounting for about 10 percent. Health impacts include both mortality and morbidity, and averting expenditures include bottled water consumption and

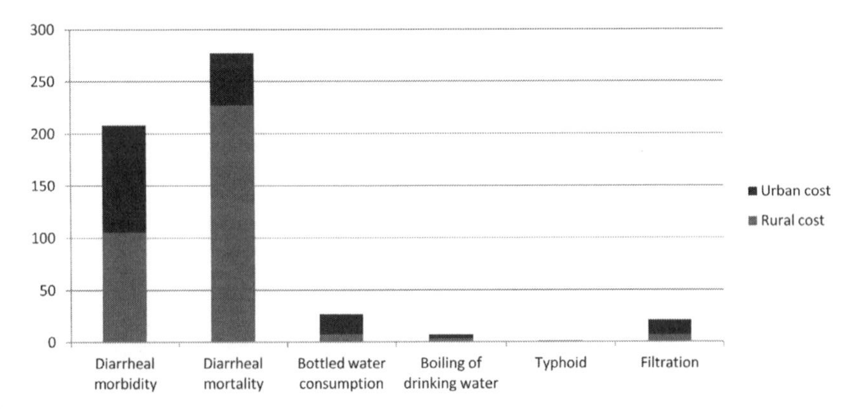

Figure 2.8 Annual costs by category (Rs billion)

Source: Staff estimates.

household boiling of drinking water. Annual costs by major category are presented in Figure 2.8.

2.5 Indoor Air Pollution

Indoor air pollution is recognized as a significant source of potential health risks to exposed populations throughout the world. The major sources of indoor air pollution worldwide include combustion of fuels, tobacco, and coal; ventilation systems; furnishings; and construction materials. The use of biomass fuel for cooking and heating in particular can give rise to indoor air pollution that threatens health, especially that of women and young children, who spend disproportionately more time indoors than do men. Several recent studies have shown strong associations between biomass fuel combustion and increased incidence of chronic bronchitis in women and acute respiratory infections in children.

The WHO (2002b) estimates that 1.6 million people die each year globally because of indoor smoke from the use of traditional fuels in the home. The most common source is incomplete combustion of fuels such as wood, agricultural residues, animal dung, charcoal, and in some countries, coal. The strongest links between indoor smoke and health are for lower respiratory infections, chronic obstructive pulmonary disease (COPD), and cancer of the respiratory system. Indoor smoke is estimated to cause, respectively, about 37.5 percent, 22 percent, and 1.5 percent of these illnesses globally (WHO, 2002b).

Firewood constitutes the major source of cooking energy in India, where more than 853 million people use firewood for cooking (Forest Survey of India, 2011). According to the 2011 census, 49 percent of the households in the country use firewood for cooking. In some states, it is as high as 80 percent. The forest-rich states have higher incidences of firewood use for cooking. There is no standard technique or operating procedure available to measure indoor air pollution. The Central Pollution Control Board (CPCB), in association with Indian Institute of Technology in Delhi, is developing a standard operating procedure on indoor air pollution in India.

There are two main steps in quantifying the health effects from indoor air pollution. First, the number of people or households exposed to pollution from solid fuels is calculated, and the extent of pollution, or concentration, is measured. Second, the health

Table 2.11 Health risks of indoor air pollution

	Odds ratios (OR)	
	"Low"	*"High"*
Acute respiratory illness (ARI)	1.9	2.7
Chronic obstructive pulmonary disease (COPD)	2.3	4.8

Source: Desaiet al. (2004).

impacts from this exposure are estimated based on epidemiological assessments. Once the health impacts are quantified, the value of this damage can be estimated.

The odds ratios in Table 2.11 have been applied to young children under the age of five years (for acute respiratory illness, or ARI) and adult females (for ARI and COPD) to estimate the increase in mortality and morbidity associated with indoor air pollution.[10] As noted, these population groups suffer the most from indoor air pollution because women spend much more of their time at home or more time cooking (with little children at their side) than older children and adult males, who spend more time outdoors.

The National Family Health Survey (2007) reports that 90 percent of rural households and 32 percent of urban households use solid fuels for cooking in India. The national weighted average is about 71 percent.

To estimate the health effects of indoor air pollution from the odds ratios in Table 2.11, baseline data for ARI and COPD need to be established. These data are presented in Table 2.12, along with unit figures for disability-adjusted life years (DALYs) lost to illness and mortality. The hypothetical level against which damages are calculated is a situation in which there is no exposure to indoor air pollution and the odds ratio is one. Some further details relating to the data are given in Appendix 1.

The results of the estimation of health losses associated with indoor air pollution are presented in Table 2.13. Estimated cases of ARI child mortality and ARI morbidity (children and female adults) from indoor air pollution represent about 38–53 percent of total ARI in India. Similarly, the estimated cases of COPD mortality and morbidity represent about 46–72 percent of total estimated female COPD from all causes.

Table 2.12 Baseline data for estimating health impacts of indoor air pollution

	Baseline		*Source*
	Urban	*Rural*	
Female COPD mortality rate (% of total female deaths)	9.5%		WHO estimate for India, Shibuya et al. (2001)
Female COPD incidence rate (per 100,000)	79		
ARI 2-week prevalence in children under 5 years	22%	22%	NFHS-3, 2006
Estimated annual cases of ARI per child under 5 years	1.0	1.0	Estimated from NFHS-3, 2006
Estimated annual cases of ARI per adult female (over 30 years old)	0.4	0.5	Estimated from a combination of NFHS-3, 2006 and Krupnick et al., 2006
ARI mortality in children under 5 years (% of child mortality)	22%		Office of Registrar General (2004)
DALYs per 100,000 cases of ARI in children under 5	165	165	Estimated from WHO tables
DALYs per 100,000 cases of ARI in female adults (over 30 years old)	700	700	
DALYs per case of ARI mortality in children under 5	34	34	
DALYs per case of COPD morbidity in adult females	2.25	2.25	
DALYs per case of COPD mortality in adult females	6	6	

Note: For details, see Appendix 1.

Table 2.13 also gives the DALYs lost to indoor air pollution. An estimated 8 million DALYs are lost each year. About 70–80 percent of these losses result from mortality and 20–30 percent from morbidity.

The central estimated costs associated with the impacts identified here are given in Table 2.14. The baseline cost data used in arriving at these estimates can be found in Appendix 1. Briefly, the cost of mortality for adults is based on the value of statistical life estimated for India as a higher bound and HCA as a lower bound for adults and on HCA for children under five. The cost

Table 2.13 Estimated annual health impacts of indoor air pollution (thousands)

	Estimated annual cases (thousands)		*Estimated annual DALYs (thousands)*	
	Urban	*Rural*	*Urban*	*Rural*
Acute respiratory illness (ARI):				
Children under the age of 5 years: increased mortality	19.5	166.4	662	5,660
Children under the age of 5 years: increased morbidity	7,570	47,925	12.5	79
Females 30 years and older: increased morbidity	9,401	47,384	65.8	331.7
Chronic obstructive pulmonary disease (COPD):				
Adult females: increased mortality	7.5	53.4	74	363
Adult females: increased morbidity	39,000	202.5	127.7	455.6
Total DALYs: mortality and morbidity			942.4	6,889.3

Source: Staff estimates.

of morbidity includes the cost of illness (medical treatment, value of lost time, etc.) and the value of DALYs estimated in GDP per capita.

To summarize, the total annual cost of indoor air pollution is estimated at Rs 305–1,425 billion, with a mean estimate of about Rs 865 billion (Table 2.14) or 1.3 percent of GDP in 2009. About 68 percent of this cost is associated with COPD, and 32 percent with ARI.[11] COPD and ARI mortality represents about 90 percent of the total cost, and morbidity about 10 percent (Figure 2.9).

Taking another classification, respiratory child mortality is 77 percent of the cost, and adult female COPD mortality is 21 percent of the cost (Figure 2.9). Acute respiratory illness in adult females and in children represents 2 percent of cost.

Health risks from indoor air pollution in household settings thus have complex linkages, and a holistic understanding of these

Table 2.14 Estimated annual cost of indoor air pollution

	Estimated annual cost (billion Rs)	
	Urban	*Rural*
Acute respiratory illness (ARI)		
Children under 5 years: increased mortality	20	190
Children under 5 years: increased morbidity	5	15
Adult females: increased morbidity	10	20
Chronic obstructive pulmonary disease (COPD)		
Adult females: increased mortality	99	485
Adult females: increased morbidity	6	15
Total	140	725

Source: Staff estimate.

linkages is crucial for the design of strategies to minimize negative impacts. The information presented here represents a small, incremental step toward better understanding of the issue of indoor air pollution exposure in the homes of rural India and has improved the evidence base for implementing and integrating environmental management initiatives in the household, energy, and health sectors.

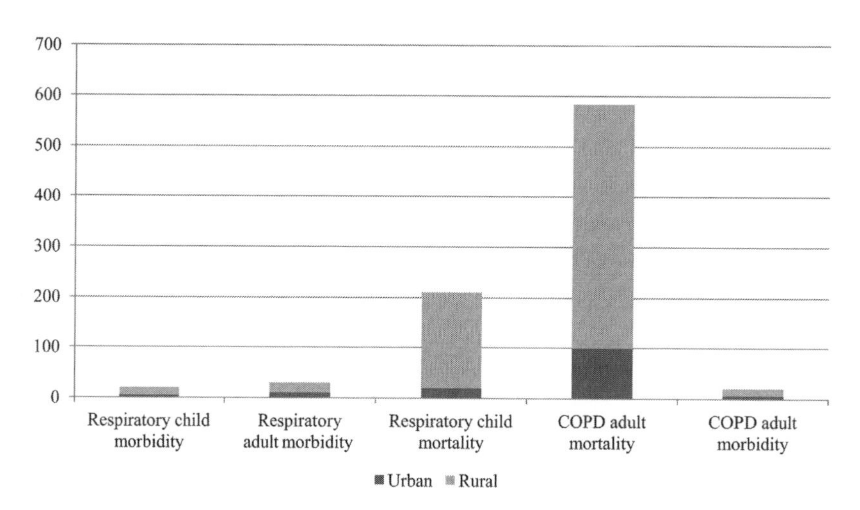

Figure 2.9 Annual costs of indoor air pollution (Rs billion)

Source: Staff estimates.

2.6 Natural Resources: Land Degradation, Crop Production, and Rangeland Degradation

As the World Bank (2007) indicates, "difficult livelihood conditions and land management practices create high dependence and pressure on local natural resources." Major categories of land degradation in India are similar to those in other Asian countries. They include (1) water and wind soil erosion and, in particular, irrigation-related land degradation, including secondary salinity, waterlogging, and soil erosion; (2) pasture and rangeland degradation; (3) degradation of forests and bushes and related loss of biodiversity; and (4) other forms of land degradation as a result of natural disasters, soil contamination, and so on. Land degradation eventually causes landslides and mudflows, especially in the sensitive mountainous areas. Most affected by degradation is pastureland near villages as well as bush and tree vegetation. Common causes are ineffective land management and lack of alternate energy resources. Land degradation not only affects agricultural productivity, biodiversity, and wildlife but also increases the likelihood for natural hazards (World Bank, 2007).

Losses to croplands and rangelands include damages from soil salinity and waterlogging caused by improper irrigation practices and human-induced soil erosion. In the absence of data on the annual increase in salinity and eroded croplands and rangelands, the annual loss of agricultural production (crop and rangeland fodder) is estimated based on accumulated degradation (see Table 2.15). This estimate may be more or less than the net present value of annual production losses depending on the rate of annual increase in degradation. The losses are considered in this section and the next.

Soil Salinity and Waterlogging

Soil salinity and waterlogging reduce the productivity of agricultural lands, and if a threshold salinity level is exceeded, the land becomes unfit for cultivation. According to conventional welfare economics, if agricultural markets are competitive, the economic costs of salinity would be measured as the losses in consumer surplus (consumer willingness to pay above market price) and producer surplus (profit) associated with the loss in productivity. These losses include direct losses through reduced yields as

Table 2.15 Land degradation in India, million hectares (2002)

Degradation type	*Degree of degradation**				
	Light	Moderate	Strong	Extreme	Total
Water erosion	27.3	111.6	5.4	4.6	148.9
a. Loss of topsoil	27.3	99.8	5.4	—	132.5
b. Terrain deterioration	—	11.8	—	4.6	16.4
Wind erosion	0.3	10.1	3.1	—	13.5
a. Loss of topsoil	0.3	5.5	0.4	—	6.2
b. Loss of topsoil/terrain deterioration	—	4.6	—	—	4.6
c. Terrain deformation/ overblowing	—	—	2.7	—	2.7
Chemical deterioration	6.5	7.3	—	—	13.8
a. Loss of nutrient	3.7	—	—	—	3.7
b. Salinization	2.8	7.3	—	—	10.1
Physical deterioration	—	—	—	—	116.6
Waterlogging	6.4	5.2	—	—	11.6
Total affected area	36.8	137.9	8.5	4.6	187.8

* Degradation is often expressed in all degradation subtypes in qualitative terms as an impact on productivity (light, moderate, strong, and extreme impact).

Source: Indiastat.com.

the land becomes saline or degraded. In practice, the calculations can be more complex because account needs to be taken of the substitution of more saline-tolerant but less profitable crops and other indirect losses. Because of a lack of data, the losses here are approximated by the value of "lost" output related to the salinity, with some simple adjustments for changes in cropping patterns.

The estimated losses from saline soils were calculated under the assumption that such land is used only for wheat production (if it is used at all). This reflects the assumption that when soils are saline, farmers will tend to plant crops that are more tolerant of this factor, such as wheat, as opposed to pulses and rice. Estimates from the United Nations Food and Agriculture Organization indicate a loss of yield of 5 percent for wheat per unit salinity (dS/m) for levels of salinity over 6 dS/m. Taking these values and applying them to lands under wheat is the basis of the estimated loss of output.[12]

The estimates indicate a net income from a hectare of land under wheat in 2009 to be in the range of Rs 8,000–18,000, and total annual losses from salinity based on the preceding assumptions are calculated at Rs 0–10 billion in scenario 1 and Rs 3–13 billion in scenario 2.[13]

In addition to these losses, we also have to account for losses from strongly saline lands that could not be cultivated at all. There are estimated to be about 13 million hectares of agricultural land that cannot be cultivated, either because the land is waterlogged or because it is highly saline. If we assume half of this area is saline, then annual net losses from land wasted due to salinity are about Rs 60–135 billion.

In total, therefore, losses due to salinity amount to between Rs 63 billion and Rs 148 billion. The middle of that range is Rs 110 billion (0.17 percent of GDP in 2010).

The losses due to waterlogging are estimated in a similar way. Then annual production losses are about Rs 20 billion or 0.03 percent of GDP in 2010.

The remaining waterlogged wasteland is estimated by Indiastat.com to be 7.5 million hectares. None of this is deemed to be cultivatable. Given that the lost annual profit for paddy production on one hectare is in the range of Rs 15,000–24,000, the annual net losses from land wasted due to waterlogging are about Rs 83–143 billion or Rs 113 billion on average (0.2 percent of GDP in 2010).

Soil Erosion

In addition to soil salinity, land degradation caused by wind and water erosion is substantial in India (Table 2.15). Two major impacts of this erosion are sedimentation of dams and loss of nutrients in the soil.

Soil erosion contributes to sedimentation of dams in India. This in turn reduces the capacity of dams and thus irrigation capacity. We do not have reliable data on sedimentation of dams and reduction in the capacity of dams in India. Hence, estimates of losses in crop production as a result of sedimentation could not be made.

Soil erosion and the loss of soil nutrients can be valued in terms of the costs of replacing the losses. The estimated cost of soil nutrients (nitrogen, phosphorus, and potassium) substitution is about Rs 320–600 billion, or Rs 460 billion on average (0.7 percent of GDP in 2010). Soil erosion is thus by far the most substantial problem of land degradation in India.

Methodology for the cost of soil salinity, waterlogging, and nutrients loss is presented in Appendix 2.

Table 2.16 Estimated annual cost of crop losses due to land degradation

	Total loss (billion Rs)			*Percent of GDP (2010)*	*Percent of GDP of the poor*
	Low	*Mean*	*High*		
Salinity losses	63	110	148	0.2%	1.1%
Waterlogged land losses	103	133	163	0.2%	1.2%
Erosion losses	320	460	600	0.7%	4.1%
Total crop land degradation losses	480	703	910	1.1%	6.4%

Source: Staff estimates.

Adding up the three categories of losses arising from land degradation in India, we get a total of Rs 703 billion or 1.1 percent of GDP in 2010 (Table 2.16). Another way to express the loss is as a percentage of GDP from agriculture, forestry, and fishery, which are sources of income predominantly for the poor, and the loss is about 6.4 percent of that GDP.

Rangeland Degradation

Land-use changes reported in India suggest that the main causes of rangeland degradation in India are irrational land-use management practices leading to denudation of vegetation from rangelands, which, exacerbated by intermittent droughts, has resulted in many pockets of desertification.[14] According to land-use data from Indiastat.com, about 10 million hectares are classified as permanent pastures. At the same time, about 1.5 times more land, including that under miscellaneous tree crops and groves and cultivable wasteland, is also used as pastures. There is a substantial share of degraded lands within all these land categories. Forest lands that are used as pastures are estimated in the next section to avoid double counting.

Open grazing in the forest is the conventional rearing practice for forest fringe communities. Overgrazing due to livestock has an adverse impact on growing stock as well as on the regeneration capacity of forest. The Indian Council of Forestry Research and Education (2001) estimates suggest that India's forests support 270 million cattle for grazing against their carrying capacity of 30 million. The incidence of grazing is estimated to affect 78 percent of India's forests,

Table 2.17 Annual cost of rangelands degradation in India

	Annual cost (billion Rs)	*Percent of GDP*	*Percent of GDP of the poor*
Market value of fodder losses	400–800	0.6–1.2%	3.6–7.2%
Foregone livestock income from fodder losses	170–256	0.3–0.4%	1.5–2.3%
Mean cost	405	0.6%	3.6%

Source: Staff estimate.

with 18 percent highly affected, 31 percent affected at medium levels, and 29 percent affected at low levels (World Bank, 2006; Ministry of Environment and Forests, 2006).

The large livestock population also results in huge collections of tree fodder, which adversely affects the forest quality. The annual requirement of dry and green fodder is estimated to be 569 and 1,025 metric tons, respectively, against the availability of 385 and 356 metric tons (Roy and Singh, 2008). This explains the pressure on India's forest from the livestock sector and the sector's contribution to the degradation of forests in human-dominated landscapes of the country. An estimated 60 percent of livestock grazes in the forest area (Kapur et al., 2010).

The loss in yield is valued in two ways. In the first method the reduction in fodder production is valued at the price of fodder. In the second method the loss of fodder is converted into a loss of livestock based on livestock feed requirements, and a value is attached to the loss of livestock. In both cases the hypothetical value against which losses are calculated is one in which original productivity prevails.

The estimated annual cost of rangeland degradation for the two methods is summarized in Table 2.17. The mean of two estimates is Rs 405 billion at 0.6 percent of GDP in 2010, or 3.6 percent GDP of the poor.

2.7 Forest Degradation

The cost of deforestation and degradation of forests is the aggregate social loss associated with degraded or deforested lands. These losses include, in theory, a wide range of local, regional, national, and even global costs. Examples include direct losses of timber, fuel wood, and nontimber products; recreation and tourism losses and indirect use losses (such as those associated with damages to ecosystem services, water supply, and carbon sequestration); and non-use value loss associated with loss of forests. This section examines each of these categories of losses with the available data.

India's forest cover is about 21 percent of total land area (about 69 million hectares). Dense forest constitutes only 12 percent of total forest cover area. Forest cover in the country has more or less stabilized since the 1980s (Nayak, Kohli, and Sharma, 2012). As per the estimates of the Forest Survey of India, forest cover increased marginally from 64.08 million hectares in 1987 to 96.2 million hectares in 2011.[15] Although forest cover area increased, the northeastern mountainous states with the densest forest, such as Nagaland, Arunachal Pradesh, Tripura, and Assam, continued to experience deforestation because of the widespread practice of shifting cultivation. With increased crop cycles and declining fallow periods in shifting cultivation practices in recent decades, the impact of traditional agricultural practice is more severe (Ravindranath et al., 2012).

Different estimates for the area under shifting cultivation range from 5 million hectares to 11.6 million hectares, involving 3 to 26 million people in 16 different states in the country (Ministry of Environment and Forests, 2006). The practice is more prominent in northeastern states. This loss is especially damaging for hilly areas, where destructive agricultural practices can result in total ecosystem destruction. Total deforested land averaged about 0.6 million hectares annually from 2006 to 2009 (indiastat.com).

Many sources reflect a substantial level of land degradation in India. Overexploitation of forest resources has led to the opening of the canopy and an increase of shrub-covered areas. The degraded area grew from 19.5 million to 24.4 million hectares in 2003 (Ministry of Environment and Forests, 2003). From the figure of 24.4 million hectares and with annual forest deforestation assumed to be at the same level as in 2006–2009, the total degraded forest area in 2009 would be estimated at 28 million hectares.

The estimated losses from the degraded forests are based on the use values attached to the forests in their nondegraded state. Previous studies have estimated the use values for two categories: direct use value and indirect use value. Under direct use value, studies have included (1) timber, (2) nontimber forest products, (3) fodder, (4) ecotourism, and (5) carbon sequestration. Under indirect use values, they have covered (1) soil erosion prevention and (2) water recharge. No estimate has been made of non-use values from forests, nor has any account been taken of biodiversity values (e.g., from bioprospecting), although these can be significant. Details of the valuation of each of these services are given in Appendix 2.

A summary of the values obtained, both in total and normalized in terms of Rs/hectare, is given in Table 2.18. The biggest source is carbon sequestration, followed by fodder and ecotourism.

In order to value the losses, we assume that degraded forests provide between 20 and 80 percent of most of the direct use values but none of the indirect values since indirect values are associated only with dense forest functions. In the case of sequestered carbon, a more precise figure is available: degraded forests are associated with 20 percent loss of total accumulated carbon (Gundimeda, 2001), reported in the range of 21–59 tons of carbon per hectare in India,[16]

Table 2.18 Estimated annual use values per hectare of forest in India (Rs billion except where indicated)

	Low	*High*
Direct		
Timber	17.2	17.2
Nontimber	21.0	21.0
Fodder	94.4	188.8
Ecotourism	51.2	51.2
Carbon sequestration	266.8	339.5
Total direct	450.6	617.7
Rs per hectare	6,471.3	8,871.2
Indirect		
Soil erosion	15.5	15.5
Water recharge	6.4	6.4
Total indirect	21.9	21.9
Rs per hectare	314.5	314.5
Total use values	472.5	639.6
Total Rs per hectare	6,785.9	9,185.7

Sources: Staff estimates applying secondary data from Gundimeda et al. (2005), Pearce et al. (1999), Ministry of Environment and Forests (2003), World Bank (2006, 2012), Ravindranath et al. (2012), Nayak, Kohli, and Sharma (2012), and data from indiastat.org and www.indg.in.

Table 2.19 Estimation of annual forest value loss (Rs per hectare, except where indicated)

Losses	*Percent loss*	*Low*	*High*
Direct values			
Timber	80–100%	198	248
Nontimber	20–100%	60	301
Fodder	0%	1,356	2,712
Ecotourism	100%	51	51
Carbon sequestration	20%	766	975
Total direct		2,432	4,287
Average percent loss		42%	53%
Total direct (billion Rs)		60.5	106.7
Indirect values			
Soil erosion	0–100%	0	1,783
Water recharge	0–100%	0	765
Total indirect		0	2,548
Average percent loss		0	100
Total indirect (billion Rs)		0.0	63.4
Total degradation losses (billion Rs)		60.5	170.2
Total deforestation losses (20% carbon losses only) (billion Rs)		9.14	25.47
Total		69.7	195.6
% GDP		0.11%	0.30%
% GDP of the poor		0.60%	1.68%

Source: Staff estimates applying secondary data from Gundimeda (2001) and Gundimeda et al. (2005)

valued at a social cost of US$20 per ton of CO_2[17] (see further explanation in Appendix 2). The losses are applied to 28 million hectares of degraded forest and about 0.6 million hectares of deforested lands.

Based on these figures, total annual losses from degraded forest land and annual deforestation losses are presented in Table 2.19. The resulting losses are in the range of 0.1–0.3 percent of GDP. We should note that this is very likely an underestimate of total losses because it excludes non-use values loss. Gundimeda et al. (2005) estimates that the non-use and bioprospecting values of forests could be as much as 6–20 times greater than use values. Because of the highly uncertain nature of this estimate, we did not use it in this study.

Notes

1. We look at damages over a relatively long period because annual figures are highly variable.
2. The difference in lower and higher bounds reflects only differences in calculation and not actual changes in losses associated with environmental degradation. A midpoint estimate presents an average of low and high estimates; the range is related to both uncertainties of valuation methods and uncertainties of exposure to specific hazards.
3. Gundimeda and Sukhdev (2008) introduced the concept "GDP of the poor," which includes GDP only from agriculture, forestry, and fishery, since these sectors reflect growth potential for most of the rural, predominantly poor Indians making up 72 percent of the total population. The importance of these sectors for the poor is also discussed in World Bank (2006).
4. The environmental media included in the analysis include outdoor and indoor air pollution; inadequate water supply, sanitation and hygiene; and natural resource degradation (soil salinity and erosion, pasture degradation, deforestation and forest degradation, and fishery loss). Losses from natural disasters were included in the COED study in Peru and in Iran.
5. Also called total suspended particulates, or TSP.
6. The focus is on particulate emissions since they are regarded as criteria pollutants and include components of other pollutants. They are an important cause of cardiovascular and pulmonary disease and lung cancer in the population, and particulate emissions levels far exceed acceptable standards in most cities.
7. Cropper et al. (2012) analyzes direct emissions from coal-burning power plants. Ambient concentrations of PM2.5 are analyzed in this report.
8. The sum of years of potential life lost due to premature mortality and the years of productive life lost due to disability (www.who.int).
9. Fewtrell and Colford (2004).
10. Although Desai, Mehta, and Smith (2004) present odds ratios for lung cancer, this effect of pollution is not estimated in this report. This is because the incidence of lung cancer among rural women is generally very low. The number of cases in rural India associated with indoor air pollution is therefore likely to be minimal.
11. Based on the mean estimated annual cost.
12. Cost of agricultural production in India is reported in Indiastat.com.
13. Information of the salinity level (slight, moderate, strong) was not available at the time of the study. Two scenarios were considered to address this issue. These scenarios are described in Appendix 2.
14. "Rangelands" is a term commonly used in World Bank studies. However, the term "grazing lands" could be substituted.
15. The forest losses considered are consistent with the findings of a number of other studies, including those published by the Forest Survey of India (FSI). We agree that in total, forest coverage in India

increased over the last decade, but we looked at degradation aspects only. We have provided conservative estimates.

16. We assume that degraded land could sequester carbon up to 80 percent of sequestration capacity of a nondegraded forest. Carbon issues are complicated, and at the next stage they should be carefully studied in the context of geographical location and other specific factors. This study attempted to provide indicative country-wide estimates.
17. The same CO_2 price is applied in *China 2030* (World Bank, 2012).

Chapter 3

How to Value?

3.1 Summary

Biodiversity underpins economic development, but it is threatened globally, and its ability to continue to provide the goods and services that support economic growth is declining.

At a global level, the implications of this have been laid out in a major report—the Millennium Ecosystem Assessment (MEA, 2005). The MEA notes that humans have made unprecedented changes to the natural world in recent decades to meet growing demands for food, fresh water, fiber, and energy and that this demand will only increase as the global population grows and consumption patterns change.

Biodiversity loss presents significant economic challenges. While a great deal of economic analysis is required to fully understand the issues, a simple and important observation is that most species and ecosystems are not traded in markets. As a result, biodiversity is under-provided. Recent interest in the economics of biodiversity and wider ecosystem services has been given empirical expression through a focus on economic valuation. Economic valuation techniques are being usefully employed to roughly estimate the value of the benefits provided by biodiversity and ecosystems.

This emphasis on economic valuation has been prompted by a growing recognition that the benefits and opportunity costs associated with biodiversity and ecosystems are frequently given only cursory consideration in policy analyses or even completely ignored. The valuation of biodiversity and ecosystem services is therefore increasingly seen as a crucial element for robust decision making, and this has been reflected in a growing body of related research. Techniques and carefully designed policy studies can assist in determining what policies are most suited to different contexts to reduce biodiversity loss cost effectively.

This topical area reports on a wide range of research that estimates the value of ecosystem services (ESS) in India, including those related to forests, grasslands, wetlands, mangroves, and coral reefs. Estimates for forest services are based on an extensive Indian green accounting study, which provides the values of timber and nontimber, fodder, forest recreation, water recharge, and prevention of soil erosion. To this we have added non-use value of forests, as well as an update of the value for forest sequestration based on the latest estimates of trends in forest management.

Estimates of the other ecosystems (grasslands, wetlands, mangroves, and coral reefs) are based on a detailed study in which all patches or sites of each such ecosystem in India were documented. Their services were valued using the characteristics of the site and a benefit function that linked the value of a site to its characteristics. This function is derived from an international database of valuation studies undertaken as part of the Economics of Ecosystems and Biodiversity (TEEB) study, and the method is referred to as benefit transfer based on meta-analysis. The valuation of each site also takes account of its mean species abundance, a measure of biodiversity, which has been estimated as part of a global research project carried out by the Netherlands Environment Agency.

The total value of services from these ecosystems is estimated to be Rs 1.4 trillion (US$29 billion) in 2009. This amounts to about 3 percent of the country's GDP in that year. The lower and upper bounds are Rs 746 billion (US$16 billion) and Rs 2.577 trillion (US$57 billion), respectively, or 1.6 to 5.5 percent of GDP.

Of the total value, forests account for 22 percent and within the forest service category, fodder is the largest. Wetlands make up the largest part of the total value (48 percent), followed by coral reefs (22 percent), grasslands (7 percent), and mangroves (2 percent).

The ESS estimates are complementary to the costs of environmental degradation (COED), which calculate the damage done to the economy and losses to the well-being of individuals as a result of the damage to the environment. The value of ecosystem services looks at the positive side of what is provided by the environment, while the COED looks at the negative side—that is, at damage caused by different types of pollution and degradation.

This study provides some important information, but further work is needed. In particular, a more accurate assessment of who benefits from these ecosystem services is important for policy purposes. While some benefits are clearly global (e.g., those from carbon sequestration), others apply to local communities (e.g., those from grasslands, wetlands, and nontimber and fodder forest services), and still others are more widely spread beyond local areas (e.g., those from timber and coral reefs). Many of these data have not yet been captured, so the estimates in this study should be taken as conservative.

In addition, some ecosystem services are missing. For example, services from lakes and rivers are important but need more data than were available. We are also missing the value of bioprospecting and some other services.

Finally, the benefit transfer method should be replaced over time with local studies of benefits, although it is impossible to cover the hundreds and thousands of patches with individual studies, and some element of benefit transfer is inevitable.

3.2 Introduction

Much has been written about loss of biodiversity in recent decades, and about the economic and social losses associated with this loss. Yet, while we have a number of pieces of anecdotal evidence, and several studies have looked at the value of biodiversity in specific contexts, no one has estimated the value of the loss of biodiversity[1] at a national or global level. This is because the links between biodiversity and biological systems and the economic and social values that they support are extremely complex. Even the measurement of biodiversity is problematic; a multidimensional metric is regarded as appropriate (Purvis and Hector, 2000; Mace, Gittleman, and Purvis, 2003) but further work is necessary to define the appropriate dimensions.

For this reason, the focus, initiated by the Millennium Ecosystem Assessment (2005), has shifted from biodiversity to measuring ecosystem services, which are related to biodiversity and are derived from the complex biophysical systems. The MEA defines ecosystem services under four headings: provisioning, regulating, cultural, and supporting, each of which has a number of subcategories.

The most important fact about these services is that they have also been facing major losses. During the last century the planet has lost 50 percent of its wetlands, 40 percent of its forests, and

35 percent of its mangroves. Around 60 percent of global ecosystem services have been degraded in just 50 years (ten Brink, 2011).

While working at the ecosystem level makes things somewhat easier, it is still important to understand the causes of the loss of these services and the links between loss of biodiversity and the loss of ecosystem services. Indeed, this is a major field of research for ecologists, and one thesis that has been developed over a long period is that more biologically diverse ecosystems are more stable and less subject to malfunction (Haines-Young and Potschin, 2010; McCann, 2000; Tilman and Downing, 1994). The links between biodiversity and ecosystem services are remain a topic of research. While some clear lines are emerging, they are not strong enough to allow a formal modeling to be carried out at a level that would produce credible estimates of the global value of biodiversity.[2]

3.3 Proposed Approach for India

Rapid economic growth has implications for use of natural capital in India, but its full value is not often factored in the context of development. Given that India is a hotspot of unique biodiversity and ecosystems, it is necessary to have a more structured approach to such valuation in the context of growth. It is, therefore, pertinent to estimate the significant contribution of natural capital in the form of ecosystem services to account or its contribution to the GDP.

The objective of this study is to obtain estimates of the current values of services from natural systems in India. To do this we, of necessity, must use ecosystem function valuation, recognizing that there is a complex link between changes in such values and the changes in the measures of biodiversity (defined appropriately). The proposed methodology takes into account (where possible) the quality of an ecosystem and the services it produces, based on the species abundance within it. This is derived from the mean species abundance (MSA) approach, which is explained more fully in section 3.5. To some extent, therefore, the study does build on the linkages between the biodiversity of a biome and its ecosystem functions.

The study quantifies and values the following ecosystem services:

1. timber services of forests
2. nontimber and fodder services of forests

3. recreation and ecotourism services of forests
4. water recharge services of forests
5. contribution of forests to prevention of soil erosion
6. carbon sequestration of forests
7. non-use values associated with dense primary forests
8. services derived from grasslands
9. services derived from wetlands
10. services derived from mangroves
11. services derived from coral reefs

The valuation of the services is based on the stock of environmental capital that exists at a point in time (as close as possible to 2010 but not always for one given year). The services are, of course, a flow from that stock. We do not value the change in ecosystem services as we do have enough data on these changes over time.

The analysis draws to the maximum extent on Indian studies.[3] This is especially the case for items 1 through 6. For the services in categories 7 through 11, the values are based on international studies but adjusted for the characteristics of the sites in India that are being considered. The procedure for deriving the estimates is to take a set of valuation functions, which calculate the value of an ecosystem per hectare as a function of national, regional, and site-specific variables. The functions were estimated as part of the TEEB study. These variables have been estimated for the individual sites in India but the meta-analytical function is estimated from international data, in which only a few Indian studies are included. There is no alternative to taking such a function, which has several limitations but is the best method currently available. In each application we indicate these limitations and point out ways in which future estimates can be improved.

This chapter should be seen as complementary to chapter 2 on the costs of environmental degradation in India. While the latter looks at the extent to which the decline in environmental services has affected the economy and well-being of the population, this chapter looks at the value of the services provided by the environmental resources that are still available to the country. As in chapter 2, all values are reported in 2009 rupees. To aid comparison, summary values are also reported in US dollars, using the exchange rate of 47.5 rupees to the dollar.

3.4 Direct and Indirect Services of Forests

The direct and indirect services of forests (apart from carbon sequestration, which is better covered by the studies referred to later) are taken from the extended study, "Green Accounting for Indian States and Union Territories Project (GAISP) 2005–2006," which was designed to build a system of adjusted national accounts for India as part of an estimate of genuine national wealth. Details of sources and the different estimates are provided in the companion to the GAISP report. Table 3.1 takes the figures from that report but excludes carbon sequestration, which has been dealt with separately.[4]

Direct use values range from Rs 184 to 278 billion (US$3.9–5.9 billion), and indirect use values amount to around Rs 22 billion (US$460 million). As noted in Gundimeda et al. (2006), these values exclude services such as bioprospecting that have not been covered due to a lack of data, but which could be significant.

Carbon Sequestration in Indian Forests

Estimates have been made of the amount of carbon stored in the forests, both in the biomass and in the soil (Kadekodi and Ravindranath, 1997; Kishwan et al., 2009). The data, which go

Table 3.1 Direct and indirect use values of forests (Rs billion unless otherwise indicated)

	Low	*High*
Direct		
Timber	17	17
Nontimber values	21	21
Fodder	94	189
Ecotourism	51	51
Total direct	184	278
Rs per hectare	6,471	8,871
Indirect		
Soil erosion	16	16
Water recharge	6	6
Total indirect	22	22
Rs per hectare	315	315

Source: Mani et al. (2012).

only to 2005, are given in Table 3.2. The Ministry of Environment and Forests (MOEF) uses these figures to make projections to 2015 under three scenarios: (1) India follows the same trend as the world average across all forests over the period 2000–2005, resulting in a decline of 0.18 percent per year; (2) India follows a trend similar to the one that prevailed domestically over the same period, resulting in an increase of 0.6 percent per year; and (3) India takes a new path that reflects the forest policy of the government, in which case forest carbon stocks grow at an annual rate of nearly 1 percent (Ministry of Environment and Forests, 2009d). Assuming these trends proceed at a constant annual rate over the period 2005–2015, we can estimate the stocks in 2009 (the year for which the estimates of the environmental flows and costs of degradation were made in the parallel study on the costs of environmental degradation in India). The corresponding stock values and annual increments to stock are given in Table 3.2.

The data show a total stock of 6.62 billion metric tons in 2005 and a possible value of between 6.57 and 6.88 billion metric tons in 2009, depending on which trend has prevailed over the period 2005–2009. Annual increments in 2009–2010 then range from a loss of 12 million tons to a gain of as much as 63 million tons.

A value of these stocks and additions to the stocks can be made based on the value of a ton of carbon. We take a range of values, with the lower bound based on the marginal costs of abatement of carbon in 2009 and the upper bound based on the social costs of carbon, as calculated across a range of studies. A review of the literature, which was conducted as part of the TEEB study, puts the lower bound at around US$8 per ton of CO_2 equivalent (Rs 380/ton), and the upper bound at around US$20.5 per ton (Rs 974/ton) (Hussain et al., 2011). The corresponding values per ton of carbon are US$29 (Rs 1,378/ton) and US$74.4 (Rs 3,534/ton). These values have been applied to the 2009 stock and flow values to give the range of estimates shown in Table 3.3. The lower bound of the stock value is between Rs 9,072 billion and Rs 9,478 billion (US$190–200 billion), and the upper bound is between Rs 23,248 billion and Rs 24,311 billion (US$490–510 billion).[5]

As a percentage of GDP, this stock has a very significant value, ranging from 19.9 to 51 percent. The annual increment to the stock could be negative over the period if world trends prevail, in which case there is a loss of between Rs 16 billion and Rs 42 billion (US$330–880 million). But with the other two scenarios, there is a

Table 3.2 Stock of carbon in Indian forests and additions to stock (millions of metric tons)

Carbon (MMT)	*1995*	*2005*	*2015*			*Stock in 2009*			*Annual Increment, 2009–2010*		
			World trends	*Past Indian trends*	*Government of India proposed trends*	*World trends*	*Past Indian trends*	*Government of India proposed trends*	*World trends*	*Past Indian trends*	*Government of India proposed trends*
In biomass	2,692	2,866	2,815	3,039	3,132	2,845	2,933	2,968	–5	17	26
In soil	3,552	3,756	3,689	3,959	4,151	3,729	3,835	3,907	–7	20	38
Total	6,245	6,622	6,504	6,998	7,283	6,574	6,768	6,875	–12	37	63

Sources: MOEF (2009); staff calculations.

Table 3.3 Values of carbon stocks in forests and increments to the stock (Rs billion)

	Stock Values		*Flow Values*	
	Lower bound	*Upper bound*	*Lower bound*	*Upper bound*
World trends	9,072	23,248	–42	–16
Past India trends	9,340	23,935	51	130
Government of India proposed trends	9,487	24,311	87	224
Average as percent of GDP	19.9	51.0	0.1	0.2

Source: Ministry of Environment and Forests (2009); staff calculations.

gain of between Rs 51 billion and Rs 87 billion (US$1.1–1.8 billion) using the lower bound of the carbon values, and between Rs 130 billion and Rs 224 billion (US$2.7–4.7 billion) using the upper bound. As a percentage of 2009 GDP, the additions to carbon amount to between 0.1 and 0.2 percent. We should also note that preventing the loss of forests, which is part of government policy, provides a benefit to the extent that any such deforestation would entail a loss of biomass equal to around 42 metric tons of carbon per hectare (Kishwan et al., 2009).

Non-use Values Associated with Conservation Forests

A number of studies show that people who have never visited a forest and have no intention of doing so are nonetheless willingness to pay to conserve forests. They draw some satisfaction from knowing that these resources are not allowed to degrade. For forests in India, both Indian citizens and citizens of other countries may have such values.

Most of the studies are for Europe and North America, although there are also a few from other countries, such as Brazil, China, Madagascar, and Israel. Based on a review of this literature, Chiabai et al. (2011) identified 27 usable estimates of passive use values for forests.[6] A meta-analysis of these studies showed that values per hectare of conserved forest varied with GDP per capita, measured in purchasing power parity (PPP) terms; with the population of the country; and with the area of conserved forest. The underlying relation is given in Appendix 3.

The source of data on forest areas as well as on the areas designated for conservation in each region is the FAO Global Forest Assessment (Food and Agriculture Organization, 2010). In the case of India, the area classified as forest under the FAO's definition was 29 percent in 2010 (Food and Agriculture Organization, 2010). According to the same source, total forest area in India in that year was 68.4 million hectares, making the area relevant for passive use valuation equal to about 19.8 million hectares.

In order to estimate the non-use values of Indian forests, we applied the meta-analysis equation to the Indian data, taking the mean and median values from the literature as the base values being transferred. The application of the above equation gives a mean and median non-use value for Indian forests of between Rs 723,947 and Rs 861,413 (US$15,241–18,135) per hectare, respectively. The corresponding total annual non-use values then amount to between Rs 14.25 billion and Rs 17.1 billion (US$300–360 million).[7]

3.5 Services from Grasslands, Wetlands, Mangroves, and Coral Reefs

The remaining categories of services are analyzed using data from a major recent study undertaken by a research group led by Salman Hussain from the Scottish Agricultural College for the United Nations Environment Program (UNEP), as part of the TEEB exercise (Hussain et al., 2011). The research combined the valuation of ecosystem services from a number of biomes with detailed GIS data on the biomes themselves. The biomes covered were forests (temperate and tropical), grasslands, wetlands, mangroves, coral reefs, and lakes and rivers. For each biome, a meta-analysis was carried out valuing the services as a function of the characteristics of the site.[8]

The estimated function was then applied to each individual site under that biome. This entailed a great deal of work. To give some indication of the level of disaggregation, the numbers of patches that were individually assessed in the benefits transfer were as follows: grasslands, 1,494,581; tropical forest, 292,822; temperate forest, 672,942; wetlands, 191,539; mangroves, 6,850; coral reefs,

16,149; and lakes and rivers, 375,316. For each patch, the team collected the biophysical data relevant to the valuation of that patch. This was done by the group based at the Netherlands Environment Agency (PBL, 2010), who kindly made the data related to patches inside the territory of India available to this team.

In separating the Indian data from the rest, some approximations were necessary. The selection of patch records in India was based on a map of the country, including islands belonging to India, as well as mangroves and coral reefs outside the mainland. The India map is produced by selecting the India+ region from the Image 24 region map (which was the basic map used in the original study) and clipping out the countries of Afghanistan, Pakistan, Nepal, Bhutan, Bangladesh, and Sri Lanka, and accompanying sliver polygons (using the standard country maps of ESRI, 2008). Coral reefs and mangroves in the original Image 24 region map were allocated a regional identification on the basis of a Euclidian allocation.

As PBL (2010) did not have accurate information about which specific coral reefs and mangroves belong to the national territory of India and which to the neighboring countries, it assumed that coral reefs or mangroves directly in front of the coast of India belonged to India and those located otherwise to the surrounding countries. It also assumed that coral reefs and mangroves around islands belonging to India also belong to India. As a consequence, coral reefs and mangroves that cross national borders were split into different parts at the location of the presumed border.

This use of benefit transfer in valuing different ecosystem services is a common procedure, although there are well-recognized limitations in terms of accuracy. A sizeable literature tests the accuracy of value transfer; Rosenberger and Stanley (2006) and Rosenberger and Johnston (2009) provide useful overviews. Although some studies find very high transfer errors (e.g., Downing and Ozuna, 1996; Kirchhoff, 1998), most studies find transfer errors in the range of 0–100 percent even in international value transfers (Ready and Navrud, 2006). Hence in the final section of this chapter, where the results are summarized, a transfer error of +/– 100 percent has been adopted.

The final remark on the valuation of ecosystem services relates to the areas to which they are applied. To account for differences in the quality of each patch, an adjustment to the area has been made based on an estimate of the mean species abundance within it. The global biodiversity model used in the TEEB study (IMAGE-GLOBIO 3) analyzes biodiversity as "the remaining mean species abundance (MSA) of original species, relative to their abundance in pristine or primary vegetation, which are assumed to be not disturbed by human activities for a prolonged period" (Alkemade et al., 2009: 375). First a relationship is estimated among a number of indicators of the pressure that an area is under and the number of species it can support relative to what it could support in natural conditions. Based on this function, an MSA value is calculated for a chosen area, given information on the different pressure indicators.

As the valuation model adopted in Hussain et al. (2011) uses the area of land as an input, the MSA value of a geographical region is calculated as the area-weighted mean of MSA values for each region. The GLOBIO 3 model is then used to assess the expected impacts of the selected drivers on MSA for a number of world regions and future scenarios, as well as the impacts of specific predefined policy measures.

The individual valuations are discussed further in the following.

Services Derived from Grasslands

Grasslands provide the following ecosystem services: food provisioning, recreation and amenity, erosion prevention, conservation of biodiversity, and carbon sequestration. An international database consisting of studies from Europe, the United States, Asia, and Africa has estimated values for all these except carbon sequestration.[9] A meta-analysis was conducted by Hussain et al. (2001) based on 19 observations. The estimated equation (see Appendix 3 for details) gives the value per hectare of grassland as a function of the grassland area within a 50 kilometer radius of the study site, the length of roads within a 50 kilometer radius of the study site, an accessibility index, and the country GDP per capita measured in PPP terms.

As noted earlier, the global database constructed by the Netherlands Environment Agency (PBL) using GIS data identified 1,494,581 patches of grassland. Of these 66,928, making up 53 million hectares, were in India. Figure 3.1 shows the location of these patches in the country.

A valuation of the services provided across the grassland patches has been made by applying the meta-analysis equation to each of

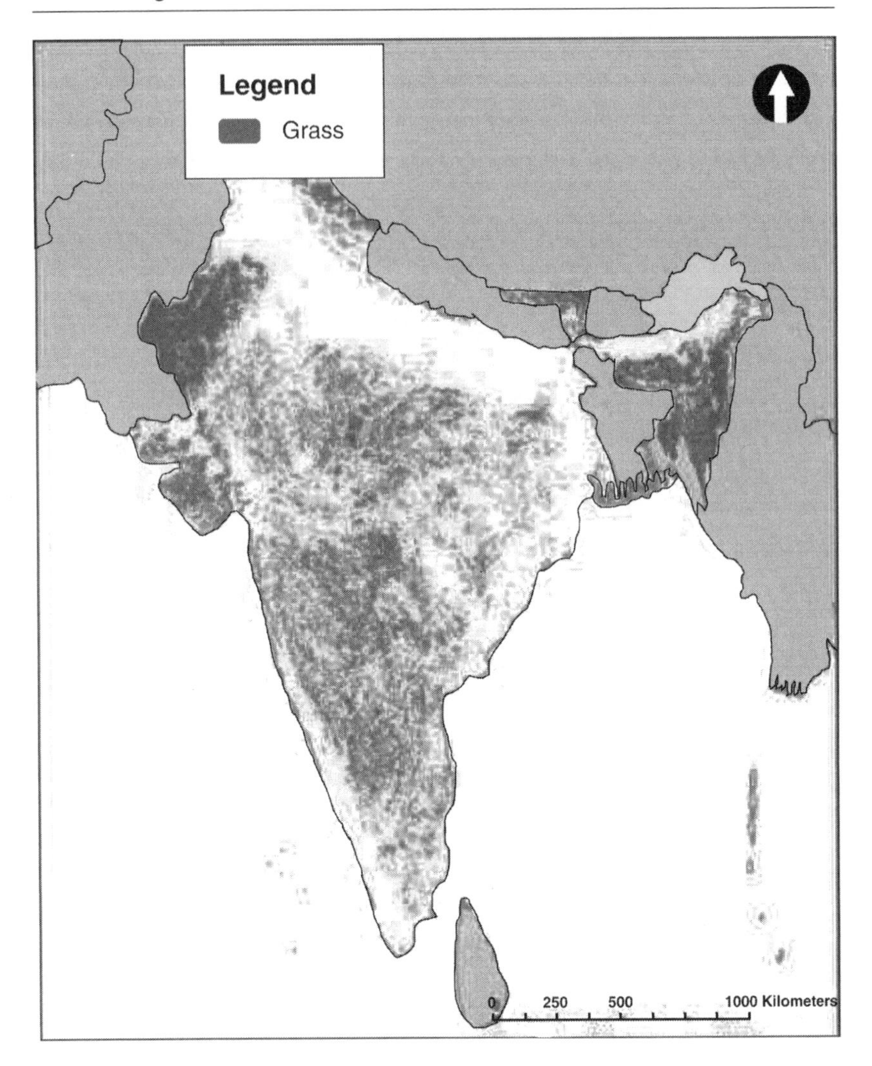

Figure 3.1 Patches of grassland in India

Source: Staff extrapolation from PBL's global database.

these patches (the PBL database provides data for the explanatory variables for each site). Naturally these vary depending on the characteristics of the patch. The range of values per hectare across the 66,928 sites is shown in Figure 3.2. The average is Rs 1,805 (US$38) per hectare, with 3 percent having a value of less than Rs 480 (US$10) per hectare and 31 percent having a value of more than Rs 2,380 (US$50) per hectare. The total value of ecosystem

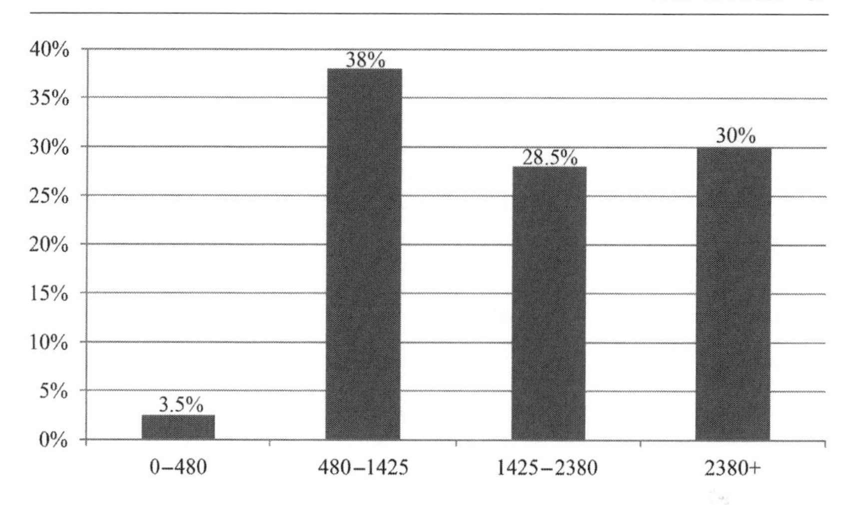

Figure 3.2 Values of grasslands in India (Rs/ha)

Source: Staff extrapolation from PBL's global database.

services from grasslands in India is estimated to be Rs 95 billion (US$2 billion) per annum.

Services Derived from Wetlands

A similar approach was taken to valuing the wetlands of India. The global meta-analysis is based on 131 studies, which generated 247 separate value estimates.[10] The studies are taken from North America, Western Europe, Southeast Asia, and Australasia. A wide range of ecosystem services are valued in these studies, including flood protection, water supply, water quality, habitat nursery, recreational hunting and fishing, food and material provisioning, fuel wood provisioning, nonconsumptive recreation, and biodiversity conservation.

The results of the meta-analysis (given in Appendix 3) found the following variables to be relevant in determining the value per hectare of wetland: the area of lakes and rivers within 50 kilometers of the site being valued, the area of wetlands within a 50 kilometer radius of the site, the population residing within 50 kilometers of the site, the length of roads within a 50 kilometer radius of the site, the human appropriation of net primary production within 50 kilometers of the site, and country GDP per capita measured in PPP.[11]

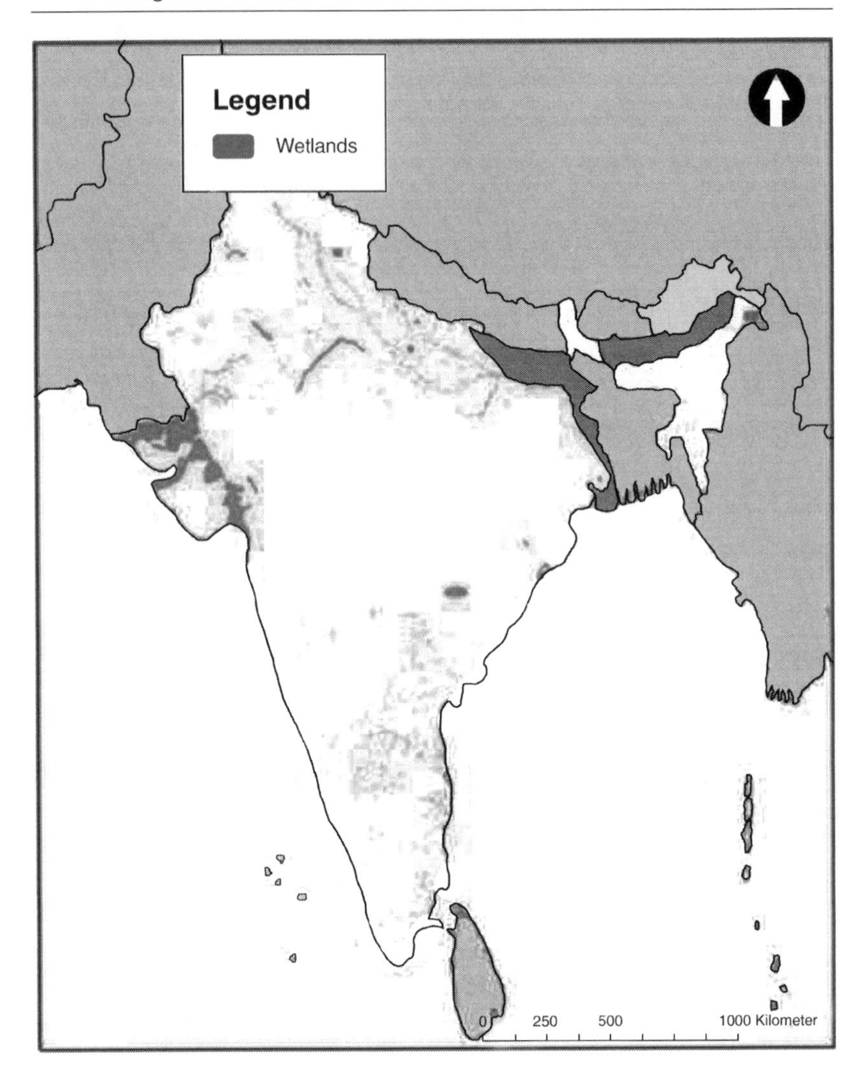

Figure 3.3 Wetlands in India

Source: Staff extrapolation from PBL's global database.

The database identified 191,539 wetlands globally, of which 3,768 are in India. Figure 3.3 shows the location of these patches, which constitute 17.5 million hectares.

By applying the meta-analysis equation to each of these patches, a valuation of the services provided across the wetlands in the country has been obtained. The range of values per hectare across the 3,768 sites is shown in Figure 3.4. The average is Rs 38,000 (US$800)

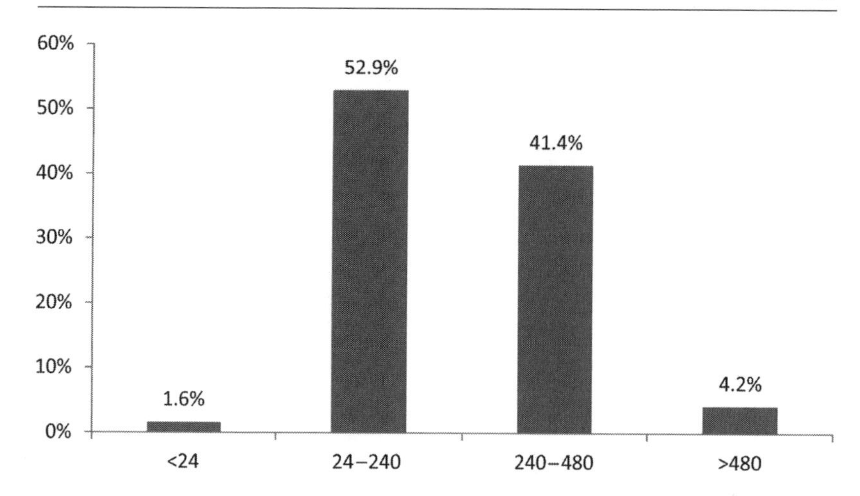

Figure 3.4 Values of wetlands in India (Rs thousand per hectare)

Source: Staff extrapolation from PBL's global database.

per hectare. Although many of the sites have a greater value per hectare, the largest sites have much lower values, so the average comes out as indicated. The total value of ecosystem services from wetlands in India is estimated to be Rs 665 billion (US$14 billion) annually.

Services Derived from Mangroves

The global database for mangroves consists of 48 original studies, from which 111 separate value estimates were obtained. These studies were conducted in Southeast Asia, Central America, the United States Gulf Coast, and East Africa. The ecosystem services included were coastal protection, water supply, water quality, habitat nursery, recreational hunting and fishing, food and material provisioning, fuel wood provisioning, nonconsumptive recreation, and biodiversity conservation.

The meta-analysis (results given in Appendix 3) found the following variables to be relevant in determining the value per hectare of mangroves: the size of the mangrove, the area of other mangroves within 50 kilometers of the site, the population resident within 50 kilometers of the site, length of roads within 50 kilometers of the site, the urban area within 50 kilometers of the site, the area of wetlands within 50 kilometers of the site, and country GDP per capita measured in PPP.

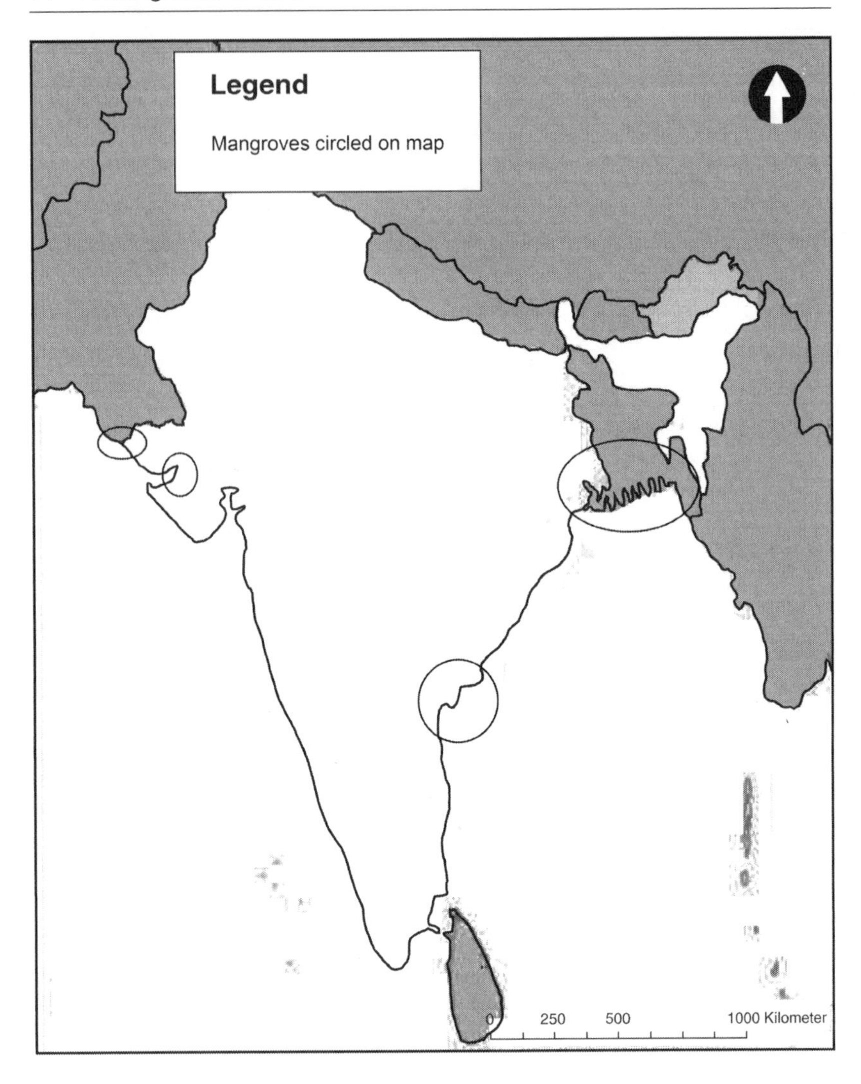

Figure 3.5 Mangroves in India

Source: Staff extrapolation from PBL's global database.

The database identified 6,850 mangroves globally, of which 89 are in India. Figure 3.5 shows the location of these patches in the country. In total they amount to 674,000 hectares. By applying the meta-analysis equation to each of these patches a valuation of the services provided across the mangroves in the country has been obtained.

The range of values per hectare across the 89 sites is shown in Figure 3.6. The average is Rs 37,860 (US$797) per hectare. The

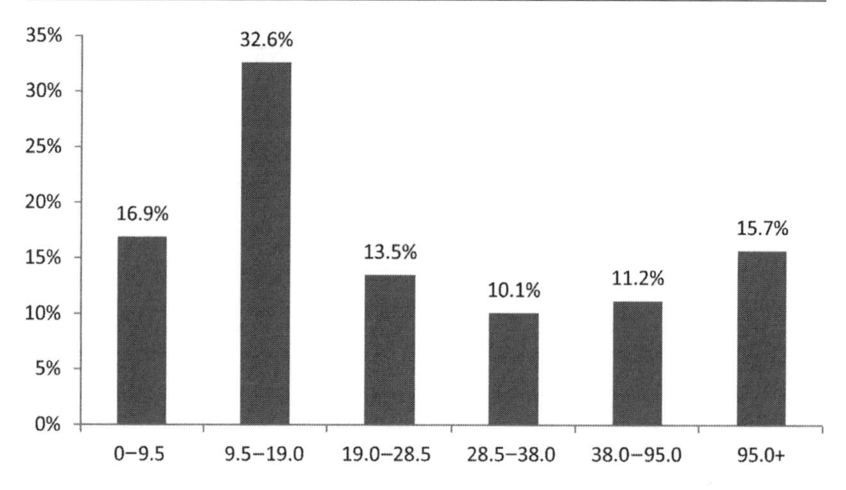

Figure 3.6 Values of mangroves in India (Rs thousands per hectare)

Source: Staff extrapolation from PBL's global database.

most common group is Rs 9,500–19,000 (US$200–400) per hectare, but around 10–12 percent of sites have values in each of the other categories: Rs 19,000–28,500 (US$400–600), Rs 28,500–38,000 (US$600–800), and Rs 38,000–95,000 (US$800–2,000) per hectare. The total value of ecosystem services from mangroves in India is estimated to be Rs 25.5 billion (US$537 million).

Services Derived from Coral Reefs

The global database for coral reefs consists of 72 original studies, from which 163 separate value estimates were obtained. These studies cover the areas with known coral reefs quite well: the Caribbean, Indian Ocean, Southeast Asia, and Pacific. The ecosystem services included were recreational diving and snorkeling, other tourism activities, recreational and commercial fishing, coastal protection, coral mining, biodiversity (including biodiversity prospecting), and non-use values.

The meta-analysis (results given in Appendix 3) found the following variables to be relevant in determining the value per hectare of coral reefs: the size of the reef, the value of income produced within 50 kilometers of the reef, the population resident within 50 kilometers of the site, length of roads within 50 kilometers of the site, the human appropriation of net primary production within 50 kilometers of the site, and the area of other coral reefs within 50 kilometers of the site.

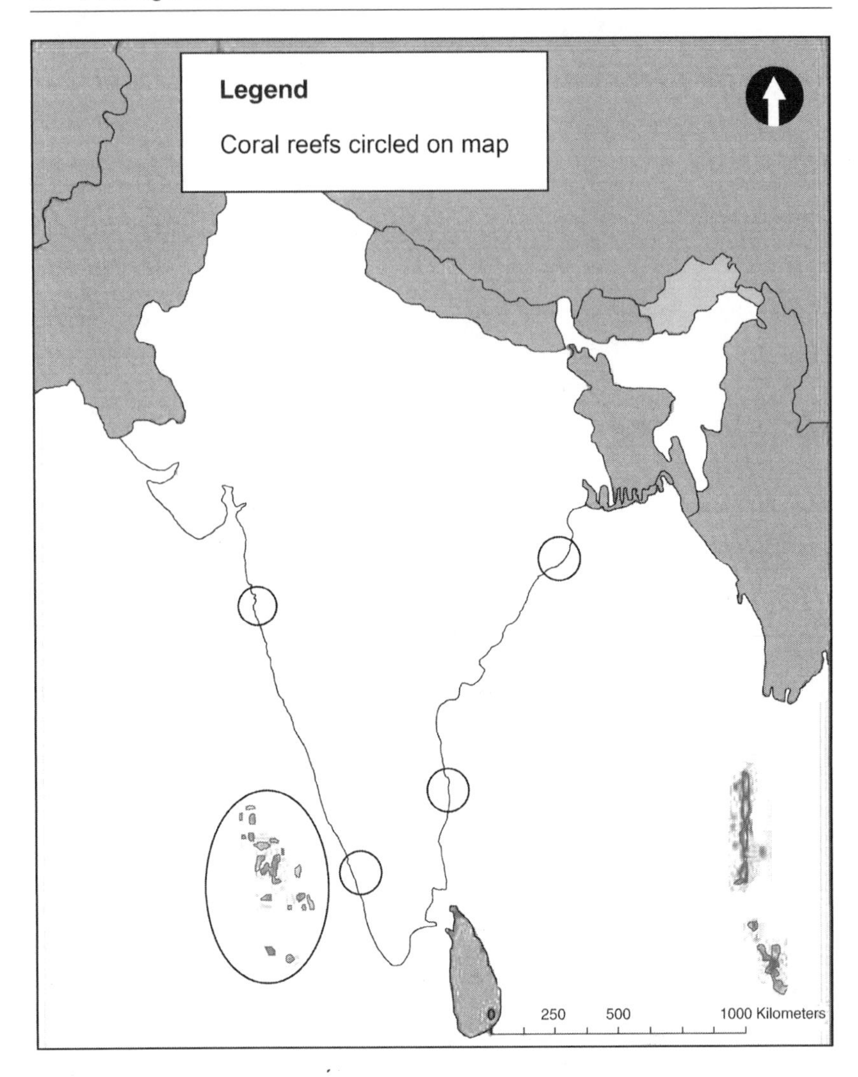

Figure 3.7 Coral reefs off the coast of India

Source: Staff extrapolation from PBL's global database.

The database identified 16,149 coral reefs globally, of which 281 were within the territory of India. Figure 3.7 shows the location of these patches in the country. They amount to 421,800 hectares. By applying the meta-analysis equation to each of these sites, a valuation of the services provided across the coral reefs in the country has been obtained.

The range of values per hectare across the 281 sites is shown in Figure 3.8. The most common range is between Rs 50,000 and

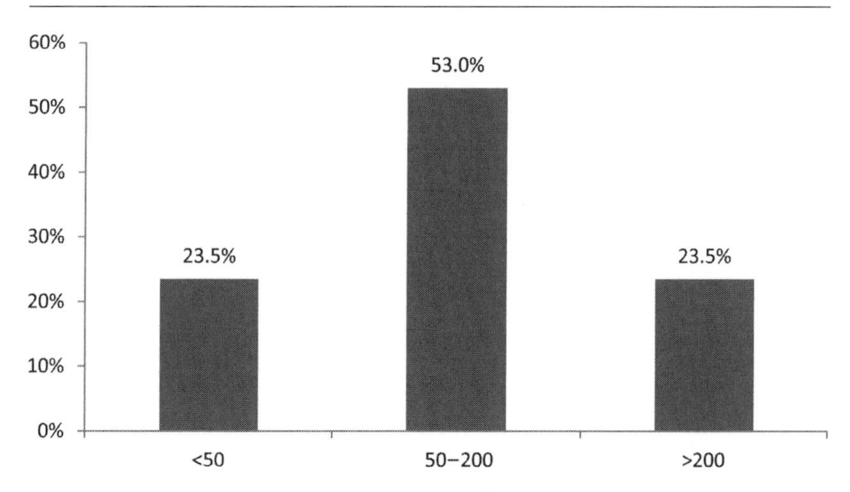

Figure 3.8 Values of coral reefs in India (Rs thousands per hectare)

Source: Staff extrapolation from PBL's global database.

Rs 200,000 (US$1,000–4,000) per hectare. Values can be as low as Rs 50,000 (US$1,000) per hectare and as high as Rs 5.7 million (US$120,000). The average is Rs 711,455 (US$15,000) per hectare. The total value of ecosystem services from coral reefs in India is estimated to be (Rs 300 billion) (US$6.3 billion).

3.6 Conclusions

This assessment has brought together the values of ecosystem services from the major biomes in India. Table 3.4 summarizes the findings and includes a range of values to allow for uncertainty in the estimates, especially those derived from the benefit transfer method. The total value amounts to about 3–5 percent of the country's GDP that year.

Some qualifications are in order to help interpret these figures:

1. The total value as a percentage of GDP central estimate of 3 percent may seem small, but this can be misleading. Another way of looking at the role of environmental resources is in terms of the GDP of the poor.[12] The services become more significant when compared with the income of the poor. The growth potential for the poor is in the share of GDP generated in agriculture, forestry, and fishery, and this share constituted about 17 percent of GDP in 2010. Although not all services from these ecosystems are for the poor, a significant percentage

Table 3.4 Values of ecosystem services in India (Rs million)

Biome/Service	*Central Value*	*Lower Bound*	*Upper Bound*	*Central as Percent of Total*
Forest carbon sequestration	40,565	−16,340	87,400	2.9%
Timber services	17,200	17,200	17,200	1.2%
Nontimber forest services	21,000	21,000	21,000	1.5%
Fodder	141,600	94,400	188,800	10.1%
Forest recreation services	51,200	51,200	51,200	3.7%
Water recharge services of forests	15,485	15,485	15,485	1.1%
Prevention of soil erosion by forests	6,413	6,413	6,413	0.5%
Non-use services of forests	15,675	14,250	17,100	1.1%
Total services from forests	309,138	203,608	404,598	22.2%
Grasslands	94,430	47,215	188,860	6.8%
Wetlands	665,950	332,975	1,331,900	47.7%
Mangroves	25,508	12,754	51,015	1.8%
Coral reefs	300,105	150,053	600,210	21.5%
Lakes and rivers	n.a.	n.a.	n.a.	n.a.
Total	1,395,131	746,605	2,576,583	100.0%

Source: Staff calculations.

of them are. If we exclude the value of carbon sequestration, coral reefs, and non-use values of forests, the remaining services amount to about 13 percent of GDP.

2. A more accurate assessment of who benefits from ecosystem services is warranted. While some benefits are clearly global (e.g., those from carbon sequestration) and others affect mainly local communities (e.g., those from grasslands, wetlands, nontimber, and fodder forest services), there are some that have different benefits for different stakeholders (e.g., those from timber and coral reefs). An attribution of how these benefits are distributed would be useful to policy makers.
3. The magnitude of the benefits of ecosystem services is less than the costs of environmental degradation, which are about 5.7 percent of GDP, as estimated in chapter 2. One reason for the difference is the major role played by damage to environmental health from air and water pollution in the latter. Arguably the

atmosphere provides provisioning ecosystem services in the form of clean air and water, but these have not been valued in the present exercise, or indeed in most estimates of the value of ecosystems. Further work is needed to improve these estimates.
4. The inclusion of services from lakes and rivers is also important, but more data are needed on household beneficiaries from lakes. We are also missing the value of bioprospecting and some other services. The benefit transfer method should be replaced over time with local studies of benefits, although it is impossible to cover the hundreds of thousands of patches with individual studies, and some use of the benefit transfer method is inevitable. Finally, the forest estimates here are based on less of a bottom-up approach than that used for the grasslands, wetlands, mangroves, and coral reefs. The forest values, however, are based on more local data, which is clearly an advantage. Eventually one would want to combine local data with the bottom-up approach.

Notes

1. For a review, see ten Brink (2011), chapter 5.4.
2. Theoretical models of the economic values attached to biodiversity have been developed. See, for example, Brock and Xepapadeas (2003). Such models draw simple links among harvesting rates, system biodiversity, and overall system value. The models, however are not yet supported by applicable empirical estimates.
3. There is an excellent report prepared by MoEF and Deutsche Gesellschaft für Internationale Zusammenarbeit (GIZ) on the economics of ecosystems and biodiversity in India (Ministry of Environment and Forests and GIZ, 2012). This report provides a very good review of the state of knowledge, challenges, valuation of ecosystem services, and biodiversity. It also suggests an approach and methodology.
4. Table 3.1 also appears in modified form as Table 7.1 in Mani et al. (2012).
5. An alternative would have been to take a value of carbon based on clean development mechanism (CDM) transactions. These values, however, are dependent on the extent to which CDM credits are allowed in Annex 1 countries under the Kyoto Protocol and can vary greatly in value. It was felt better to stay with the valuations of carbon used in international studies.
6. The review covered sources such as EconLit, EVRI database, and IUCN database for forest studies.
7. All original study values were converted into 2009 dollars and converted to rupees at an exchange rate of 47.5 rupees to the dollar.

8. In all cases but one (lakes and rivers), values were estimated in US dollars per hectare. For lakes and rivers, the meta-analysis was in terms of willingness to pay per household.
9. Contingent valuation and choice experiments have been used for recreational values of grasslands and wildlife conservation; hedonic pricing has been used for the amenity value; and net factor income and market prices have been used to estimate food provisioning (Hussain et al., 2011).
10. The collection of wetlands in the database used included coastal wetlands but not mangroves, which are treated separately. The data was provided by PBL. A reviewer from the Ministry of Environment and Forests has noted there are differences between this set and the wetlands identified in the The National Wetland Atlas prepared by the Space Application Center in 2011. The latter for example has only 15.25 million hectares of wetlands, whereas our database has 17.5 million hectares. We have stayed with our database because we did not have the individual patches in the Wetland Atlas, nor the defining variables for the meta-analysis.
11. Note that the database does not include lakes and rivers as part of the wetland ecosystem, although these bodies affect the unit value of a given wetland. The choice of regressors indicates that the valuation model is more focused on in situ conditions and does not address the scale dimension of ecosystem functioning. This is a limitation of the analysis and something to be addressed in future work.
12. Gundimeda and Sukhdev (2008) introduced the concept of "GDP of the poor." It includes GDP only from agriculture, forestry, and fishery, since these sectors reflect the most growth potential for the rural poor in India, who comprise 72 percent of the total poor. Of course, not all income in these sectors goes to the poor, but possibly a higher percentage does than in most other sectors. A reviewer from MOEF has noted that while GDP of the poor is a useful theoretic construct, a meaningful analysis would have been to spatially explore the correlations between incidence of poverty and ecosystem service values. This can then provide better insights into conservation targeting and prioritization. This is something to plan in future work.

Chapter 4

What Are the Trade-Offs?

4.1 Summary

One of the key environmental problems facing India is that of particle pollution from the combustion of fossil fuels. This has serious health consequences and with the rapid growth in the economy these impacts are increasing. At the same time economic growth is also imperative and policy makers are concerned about the possibility that pollution reduction measures could significantly impede that growth.

This topical area addresses the trade-offs involved in controlling local pollutants such as particles. Using an established computable general equilibrium (CGE) model, it evaluates the economic impacts that would result in targeted PM10 emission levels 10-percent and 30-percent lower, respectively, than they otherwise would be in 2030. The impacts examined are changes in GDP, health indicators, and emissions of CO_2.

Corresponding to the 10-percent PM10 reduction target a "green growth" scenario is projected, and corresponding to the 30-percent target a "green growth plus" scenario is envisaged. Each scenario is compared to a business-as-usual (BAU) scenario. In the green growth scenario the taxes induce a shift to a greener fuel mix and modest annual energy efficiency gains over and above the historic trend. In the green growth plus scenario the same taxes result in a greater improvement in the performance of coal technologies and in larger reductions of particle emissions per unit of output throughout the economy through modernization and renovation of existing capital. In both scenarios these changes work partly through the increases in the prices of fuels that generate particles and partly through the pressures that the taxes generate for wider improvements in energy efficiency. The latter was introduced exogenously in the model. There is considerable evidence in the literature that such exogenous improvements do come about when economic instruments such as environmental taxes are introduced.

The main findings are as follows:

1. A 10-percent reduction in particulate emission results in a lower GDP but the size of the reduction is modest. With a PM10 tax, conventional GDP would be about US$46 billion lower in 2030, representing a loss in growth of 0.3 percent with respect to BAU. The impact on GDP is greater (0.5 percent) if we seek to achieve the PM reduction target via a coal tax. The impact on GDP growth rate is almost negligible when compared to BAU in both cases.
2. For a 30-percent reduction in particulate emission, conventional GDP is about US$97 billion lower in 2030, representing a loss of 0.7 percent with a PM10 tax. Thus the scenario suggests that even a substantial reduction in emissions can be achieved without compromising much on GDP growth rates if it can be supported by adequate least-cost policy measures. Again the coal tax performs worse, with a GDP loss of 1.07 percent.
3. These losses in GDP from the tax are partly offset by the health gains from lower particle emissions. In the case of a 10-percent reduction, the central estimate of gains in the form of less premature mortality and reduced cases of respiratory and cardiovascular diseases is US$35 billion in 2030; in the case of a 30-percent reduction these gains amount to US$67 billion as a central estimate.
4. In addition the taxes reduce emissions of CO_2 by about 590 million metric tons in 2030 in the case of the 10-percent reduction in PM10 and 830 million metric tons in the case of the 30-percent reduction in PM10. Valuing such reductions at a modest US$10 per ton would yield a benefit that by itself would cancel out the loss of GDP in the 10-percent case and be slightly less than the loss of GDP in the 30-percent case.
5. Taken together, the gains from CO_2 reduction and the health benefits are greater than the loss of GDP in both cases.

The results thus support a reform based on fiscal measures that address the problems of particle pollution. There are, however, several issues that need further investigation. In particular, the distributional implications of the proposed taxes need to be analyzed and the mechanisms by which environmental taxes motivate improvements in energy efficiency need to be better understood.

4.2 Economic Growth and Environmental Sustainability: What Are the Trade-Offs?

This topical area analyzes some of the key trade-offs between economic growth and environmental sustainability for India. The tool used for this analysis is a computable general equilibrium (CGE) model.[1] CGE models are powerful tools for tracing how changes in one sector are propagated through the rest of the economy, affecting dependent sectors, patterns of trade, income and consumption, and the fiscal and international financing needed for macroeconomic stability and growth goals (see Figure 4.1).

CGE models are widely used to analyze the aggregate welfare and distributional impacts of policies whose effects may be transmitted through multiple markets. They can also be deployed to analyze the effects of specific instruments or a combination of instruments. Examples of their application may be found in areas as diverse as fiscal reform and development planning (see, for example, Perry, Whalley, and McMahon, 2001; Gunning and Keyzer, 1995), international trade (Shields and Francois, 1994; Martin

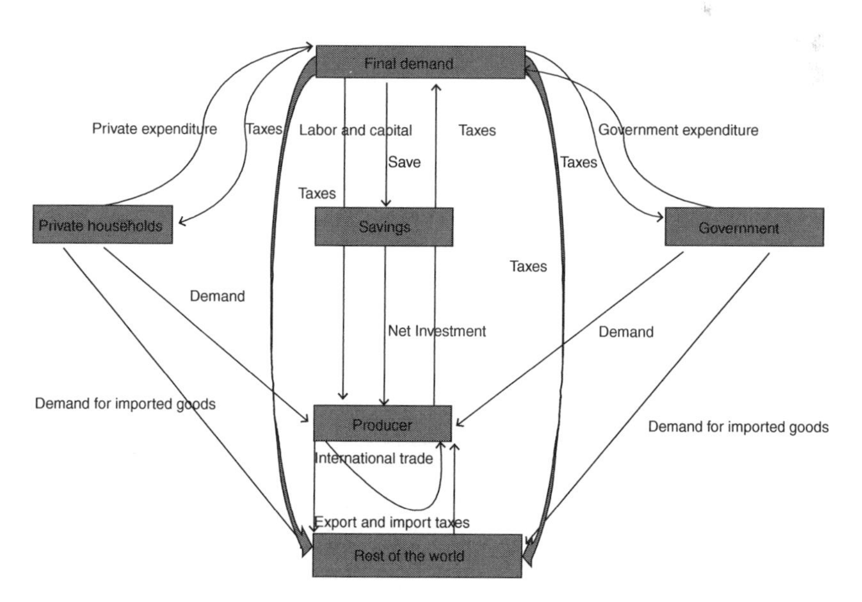

Figure 4.1 Description of CGE model

and Winters, 1996; Harrison, Rutherford, and Tarr, 1997), and increasingly environmental regulation (Weyant, 1999; Bovebourg and Goulder, 1996; Goulder, 2002) (see Box 4.1).

The CGE model used here is based on a framework developed and maintained by the Global Trade Analysis Project (GTAP)

Box 4.1 CGE Models and Environmental Policy

Policies aimed at significantly reducing environmental problems such as global warming, acid rain, deforestation, waste disposal, or any degradation in air, water, or soil quality may imply costs in terms of lower growth of GDP, a reduction in international competitiveness, or a reduction in employment. The implied change in relative prices will induce general equilibrium effects throughout the economy. For this reason, it is useful to evaluate the effects of environmental policy measures within the framework of a CGE model. Although partial equilibrium models make it possible to estimate the costs of environmental policy measures, taking substitution processes in production and consumption as well as market clearing conditions into account, CGE models additionally allow for adjustments in all sectors, enabling us to consider the interactions between the intermediate input market and markets for other commodities or intermediate inputs, thereby completing the link between factor incomes and consumer expenditure.

Since the first environmental CGE models appeared (Forsund and Storm, 1988; Dufournaud, Harrington, and Rogers, 1988), the literature has included applications in many major areas, such as (a) models used to evaluate the effects of trade policies or international trade agreements on the environment (Lucas, Wheeler, and Hettige, 1992; Grossman and Krueger, 1993; Madrid-Aris, 1998; Yang, 2001; Beghin et al., 2002) and for diverse applications in the area of the Global Trade Analysis Project (Hertel et al., 1997); (b) models to evaluate climate change, which are usually focused on the stabilization of CO_2, NO_x, and SO_x emissions (Bergman, 1991; Jorgenson and Wilcoxen, 1993; Edwards and Hutton, 2001); (c) models focused on energy issues, which usually apply energy taxation or pricing to evaluate the impacts that changes in the price of energy can have on pollution or costs control (Pigott, Whalley, and Wigle, 1992); (d) natural resource allocation or management models, whose objective is usually the efficient interregional or intersectoral allocation of multi-use natural resources—for example, allocation of water resources among agriculture, mining, industry, tourism, human consumption, and ecological watersheds (Robinson and Gelhar, 1995; Ianchovichina, Roy, and Shoemaker, 2001); and (e) models focused on evaluating the economic impacts of environmental instruments or of specific environmental regulations, such as the US Clean Air Act (Jorgenson and Wilcoxen, 1990; Hazilla and Koop, 1990).

The CGE modeling in India with environmental links has mainly focused on reduction of carbon emissions and its implications for economic growth (Murthy, Panda, and Parikh, 2000; Ojha, 2005, 2008). With a view to develop a fact-based perspective on climate change in India, the Ministry of Environment and Forests has supported a set of independent studies by leading economic institutions. This initiative is aimed at better reflecting the policy and regulatory structure in India and its specific climate change vulnerabilities. The studies, which use distinct methodologies, are based on the development of energy-economic and impact models that enable an integrated assessment of India's greenhouse gas (GHG) emissions profile, mitigation options, and costs, as well as the economic and food security implications (Ministry of Environment, 2009).

network.[2] The GTAP model is built on a global trade database and reflects, among other indicators, India's performance in terms of export growth, which has increased dramatically during the last decade. With India emerging as a major producer and exporter of goods, including pollution-intensive commodities, the use of such a model to assess the environmental impacts of the country's development path was considered appropriate. The main environmental variable that has been included in the model is emissions of particulate matter of less than ten microns (PM10), as well as particles of sulfates and nitrates. These emissions are recognized as among the most important in terms of their health effects.

The standard GTAP model has been expanded to include emissions from all the key sectors, including PM10 and other small particles emissions originating from fuel use and production activities. A detailed description of the model, assumptions, and corresponding equations is given in Appendix 4.

This is the first time that a CGE model for India has looked at the trade-offs between economic growth and "local" pollution mitigation.[3] The open economy model incorporates links among 57 sectors within agriculture, manufacturing, and services, as well as links between the economic output of these sectors and air pollution emissions, principally PM10 and emissions of SO_2 and NO_x which give rise to health effects. Other CGE models for India have so far included only 11 to 36 sectors and have not tracked emissions such as PM10.

The model's database developed by the GTAP network[4] (GTAP database version 8 for 2007) includes data from India's National Accounts. This was complemented with statistics on urban pollutants (from national statistical sources) and macroeconomic variables (i.e., growth rate projections and total factor productivity [TFP] from the literature). Specifically, the model was extended by several external inputs, such as demographics, labor productivity, labor supply, and emission coefficients.

4.3 Methodology

First, an economic growth scenario was developed, reflecting the most likely path that the Indian economy could follow from 2010 through 2030. This path represents the "economic baseline." The GTAP model was calibrated to reproduce actual GDP growth rates in the country during 2007–2010 and growth projections in line with World Economic Outlook projections.[5] While the recent IMF survey of the Indian economy suggests a robust 7–8 percent growth in the next few years in spite of a global economic slowdown, it will be necessary, according to the IMF, to focus on reinvigorating the structural agenda, rather than relying on monetary and fiscal stimulus to ensure sustainable growth. Measures to facilitate infrastructure investment, reform of the financial sector and labor markets, and measures to address agricultural productivity and skills mismatches stand out. Also according to IMF, reorienting expenditure toward social areas is vital to make growth more inclusive (which, in turn, would boost growth).[6]

Second, an "environmental baseline" was constructed according to our estimations of PM10 and other small particle emissions.[7] Third, a health module was developed outside the CGE to estimate the health impacts expected to occur during the same period: the potential mortality and morbidity effects of such small particles.[8] The pollution impact on health is characterized by mortality and morbidity figures for three different pollution scenarios ("upper," "central," and "lower").[9] These reflect the uncertainties about the magnitude of the impacts of PM10 and other small particles.

The main analysis carried out was to evaluate the economic and environmental impacts of a 10-percent reduction or a 30-percent reduction in PM10 and other small particle emissions relative to what they would be in 2030 under a business-as-usual scenario. To achieve these targets, two different types of policy instruments in addition to an increase in autonomous energy efficiency and investment in clean energy were considered:

1. a tax on coal alone; or
2. a tax on PM10 and other small particles, translated into a tax on the fuels that generate PM10, namely coal and oil.[10]

In each case, the model was run to look at the effects of the taxes on conventional GDP and their impacts on particulate emissions. The health damages are dealt with outside of the model.

The application of tax policies in the model should not be construed as an endorsement of these specific policy approaches. Tax policies are an analytically convenient way to represent a broader class of policies that use economic incentives to change behavior, including an emissions trading system. However, our approach can less readily be interpreted as showing the impacts of more prescriptive emission control policies, such as specific technology standards, which generally are costlier—sometimes much more so—than incentive-based policies. Moreover, the CGE approach has limitations in its ability to fully reflect the potential for "low hanging fruits," notably improvements in thermal and end-use energy efficiency that can yield reduced emissions as a co-benefit (i.e., between CO_2 and PM10). This point plays an important part in our analysis, as described in later sections.[11]

In terms of environmental impacts, the model was expanded to estimate PM10 emissions and generation of sulfates and nitrates of similar diameter through the year 2030 based on fuel use and production. These pollutants are the most important of all air pollutants in terms of their health impacts and are associated with significant additional mortality and morbidity for the population, including the labor force (see Box 4.2). In this study, morbidity was quantified by estimating the days lost due to reduced activities and increased hospital admissions due to respiratory illnesses. Each of these impacts was quantified based on epidemiological studies (more details are in Appendix 4). Based on the CGE model estimation of emissions, the increase in PM10 and other small particle concentrations was estimated using the concept of uniform rollback.[12] Under this assumption, health impacts can be linked directly to levels of emissions; the analysis does not include a characterization of how emissions affect air quality (pollutant concentration), the measure one would typically see in the health literature to estimate changes in illness and risk of premature death.

The morbidity and premature mortality impacts of PM concentrations were measured in monetary terms as follows. For morbidity, an estimate was made of losses in productivity and costs of treatment for illness. For premature mortality the impacts were valued in terms of both loss of future productivity (where appropriate) and the welfare loss associated with early death (see Appendix 4, section 5 for details).

It is often the case that if an environmental policy such as a tax induces technical change, for example by triggering emission reduction or resource-saving technical change, it reduces the cost of achieving a given abatement or resource conservation target. For example, emission of air pollutants can be cost effectively reduced by fuel substitution (nonenergy for energy or within-energy inputs) and by efficiency improvements in power generation and use. Most CGE models, however, assume no difference in the pattern of technical change between the base case and the policy case, which often leads to an upward bias in the cost estimate of policy. Other common approaches to take into account with technical change are the use of capital vintages involving different technologies or the modeling of autonomous energy efficiency improvements. An attempt is therefore made in the CGE model to capture these technological shifts over time by altering the elasticity of substitution between capital and energy and by altering levels and types of investments and corresponding emission coefficients (in line with the existing bottom-up analyses for India). These are described in detail in Appendix 4.

Box 4.2 Particulate Emissions in India

Particulate matter is by far the most problematic air pollutant on a national scale, with annual average concentrations of suspended particulate matter (SPM) exceeding the National Ambient Air Quality Standards (NAAQS) in most cities (Central Pollution Control Board, 2006; Ministry of Environment and Forests, 2009). India's national average of 206.7$\mu m/m^3$ SPM in 2007 was well above the old NAAQS of 140 $\mu g/m^3$ for residential areas. Most Indian cities exceed, sometimes dramatically, the current NAAQS of 60 $\mu m/m^3$ for respirable suspended particulate matter (RSPM). Average annual concentration of RSPM in Delhi for example is about 120 $\mu g/m^3$, as compared to the residential NAAQS of 60 $\mu g/m^3$ and World Health Organization (WHO) guidelines of 20 $\mu g/m^3$ (Central Pollution Control Board, 2006; WHO, 2008). Five of six cities covered in a recent report exceeded the standard in all years 2000–2006 (Central Pollution Control Board, 2011). By contrast, sulfur dioxide (SO_2) and nitrogen oxides (NO_X) are less of a problem in India. Most cities are below the NAAQS for these pollutants.

The figures refer to both SPM and RSPM. SPM is a broader category referring to all suspended particulate matter of less than 100 micrometers in diameter. Research on the health effects of particulate matter indicates that the smaller particles in RSPM are more dangerous for health because they penetrate more deeply into the lungs (U.S. Environmental Protection Agency, Climate Change Division, 2008). In India, RSPM is defined as fine particles less than 10 μm (PM10). Other countries refer to this pollutant as PM10 and may also measure PM2.5 (i.e., particles of less than 2.5 μm in diameter).

Indian standards recognize the danger of air pollution. In November 2009, the Ministry of Environment and Forests (MoEF) announced new NAAQS (Central Pollution Control Board, 2009). Compared to the previous version from 1994, the revised NAAQS brought six new pollutants under regulation (including introducing a standard for PM2.5), tightened the acceptable ambient concentration for other pollutants, and eliminated the distinction between industrial and residential areas. As a result, many urban areas—which may have been out of compliance even with the older norms—must significantly cut emissions to move toward the more stringent, uniform standards now in place. The shift from regulation of ambient SPM to RSPM in the new NAAQS in particular is significant in directing the focus of regulation to those pollutants that matter for human health. India's MoEF has launched a pilot emissions trading scheme in three states to improve air quality and help the states meet the new NAAQS.

Source: Adapted from Greenstone et al. (2012).

4.4 Scenarios

As noted above the model was run for the business-as-usual scenario, plus six scenarios reflecting a menu of instruments that look at the impacts of reducing PM10 and other particles through different tax instruments (see Figure 4.2). Details of the different scenarios are given in Table 4.1.

Two types of taxes are modeled. A domestic fuel tax (*to*) is added to the producer price (*ps*) for coal, oil, natural gas, and refined oil to obtain the market price (*pm*):

$$pm(i,r,t) = to(i,r,t) + ps(i,r,t) \tag{1}$$

$to(i,r,t) > 0$

$ps(i,r,t)$: producer price for commodity (i) in region (r) in year (t)

$to(i,r,t)$: tax on commodity (i) in region (r) in year (t)

$pm(i,r,t)$: market price of commodity (i) in region (r) in year (t)

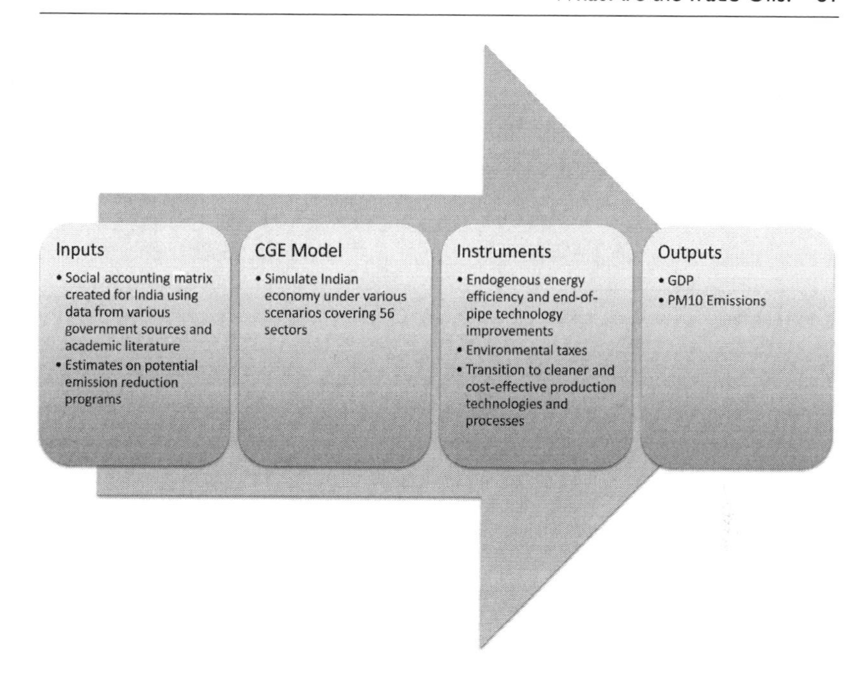

Figure 4.2 How the CGE model works

The tax rate increases linearly at a constant rate starting from 2012. Tax rates (*to*) used in different scenarios are displayed in Table 4.2.

$$to(i,r,t) = to(i,r,t-1) \times \Delta to(i,r,t) \qquad (2)$$

An imported fuel tax (*tm*) is applied to the import price (*pms*) of coal, oil, natural gas, and refined oil. The change in *to* is strictly positive, decreasing over time.

$$pms(i,r,t) = tm(i,r,t) + tms(i,r,s,t) + pcif(i,r,s,t) \qquad (3)$$

pms(*i*,*r*,*t*): import price of commodity (*i*) by region (*r*) in year (*t*)

tms(*i*,*r*,*s*,*t*): ad valorem tariff on commodity (*i*) in region (*r*) imported from region (*s*) in year (*t*)

tm(*i*,*r*,*t*): tax on import of (*i*) from region (*r*) in year (*t*)

pcif(*i*,*r*,*s*,*t*): border price of commodity (*i*) in region (*r*) imported from region (*s*) in year (*t*)

Table 4.1 CGE model: scenarios

Scenarios	*Instruments*	*Assumptions*	*Outcomes*
Business-as-usual GDP growth		Economic growth of approximately 7 percent per annum.	Some PM emission reduction because of increase in autonomous energy efficiency of supply and end-use technologies (driven by current policies).
Green growth	Using a tax on coal only. Tax applied to both domestic and imported coal. Using a tax on PM10. Tax applied to coal and oil in relation to the emissions of PM10 and other small particles	Tax-induced shift to a greener fuel mix and annual energy efficiency gains over and above the historic trend. Limited investment availability and turnover of capital stock.	A 10-percent reduction in PM10 and other small particles in 2030 over and above reductions achieved under business-as-usual.
Green growth plus	Using a tax on coal only. Tax applied to both domestic and imported coal. Using a tax on PM10. Tax applied to coal and oil in relation to the emissions of PM10 and other small particles.	Tax-induced shift leading to significant improvement in coal technologies along with change in plant vintages over time. Higher investment availability and faster turnover of capital stock.	A 30-percent reduction in PM10 and other small particles in 2030 over and above reductions achieved under business-as-usual.

The change in $tms(i,r,t)$ is strictly positive, decreasing over time:

$$tm(i,r,t) = tm(t,r,t-1) \times \Delta tms(i,r,t) \quad (4)$$

Table 4.2 Different taxes (%) for a 10-percent reduction in PM emissions by 2030—derived from the CGE simulations

Tax Regime	*Applied to*	*2014*	*2030*
Coal tax	Coal	14.0%	38.5%
PM tax	Coal, oil	3.4%	16.2%

The tax rate on imported fuel also increases linearly at a constant rate starting from 2012. Tax rates used in different scenarios are given in Table 4.2.

Business-as-Usual GDP Growth Scenario

The business-as-usual GDP growth scenario refers to a purely economic baseline and is based on past economic performance for 2007–2010 and on IMF projections of GDP for 2011–2015, with associated projections up to 2030 derived from projections for TFP. The model then calculates the required investments to achieve the projected growth, along with the demands for different types of fuel. Domestic prices for fuel as well as other goods are determined so that demand and supply are equated. Some emission reduction (and therefore decline in PM intensity of GDP) happens under the business-as-usual scenario due to autonomous technological change built into the model.[13] This is partly driven by the macroeconomic structural shift away from the agriculture sector toward knowledge-based industries; greater and easier access to global knowledge, technology, and capital; and the growth impetus provided by the commercial and services sectors. In addition the shift also reflects the recent policy initiatives to reduce the sulfur content of diesel in the transport sector, the use of compressed natural gas for public transport, emissions limiting performance standards for passenger vehicles, and stricter enforcement of existing environmental laws.[14]

Green Growth Scenario

The green growth scenario targets a reduction in PM10 and other small emissions by 10 percent more than what could be achieved relative to business-as-usual in 2030. The green growth scenario is thus a modified version of the business-as-usual scenario in which a tax instrument is used to achieve a targeted emissions reduction.

It is important to note that the green growth scenario as mentioned here is in the narrower context of incentivizing the private sector. Broader concepts of green growth are usually defined by sustainable development options and challenges of social and economic development. Green growth scenario is modeled through a tax on coal or PM10.[15] A tax on polluting inputs will raise the unit cost of production, and, responding to the rise in unit cost of production,[16] the producer will reduce the output or substitute it with a more ecofriendly input. Either of these actions will reduce pollution. It is thus anticipated that the tax in the model will encourage a shift to a greener fuel mix and annual energy efficiency gains over and above the historic trend. In the case of a tax on PM10, for instance, we consider a modest tax as a way of reducing particle emissions per unit of coal used. For further reductions in PM the tax has to induce a shift out of coal to cleaner fuel. The green growth scenario is summarized in Table 4.1.

Green Growth Plus Scenario

The green growth plus scenario incorporates a more aggressive target of a 30-percent reduction in PM10 and other small particles in the air by 2030 over what could be achieved under the business-as-usual scenario. Here again, targeted small particles emissions reduction is attained through a tax on coal or PM10.

One important difference between the green growth and green growth plus scenarios is that the latter assumes that, as the economy matures, the market realizes the economic benefits of cleaner and more efficient production. Gradually the environmental command-and-control "push" policies in the initial periods are replaced in the medium to long run by market-driven "pull" policies to achieve cleaner and more efficient production. For example, the performance of coal technologies improves over time, reflected in their rising plant load factor and growth of newer plant vintages, with more of the older, less-efficient plants getting replaced, and in the increased penetration of advanced coal technologies like supercritical pulverized coal and integrated gasification combined cycle, which become competitive over time. While recognizing the limitations of incorporating all these technologies within the CGE framework, they have been modeled through broad alterations in investments and emission coefficients. The idea is that the latest vintage, added to aggregate capital stock, embodies innovation and technological improvement with no additional cost to the producer.[17]

The CGE model used in this analysis was limited in terms of formulating different policy scenarios because the current dataset included only five types of energy sources: coal, crude oil, refined oil and coal products, natural gas, and electricity. Based on data availability, the model and study can be expanded in the future to include other energy sources, such as renewable energy and carbon sequestration measures.

Calibrating the Model for the Business-as-Usual GDP Growth Scenario

Estimates of growth in population and labor force were based on projections made by national and international sources (e.g., National Council for Applied Economic Research [NCAER], United Nations, and World Bank). Medium projections were used for measuring population growth in 2007–2030 using UN demographic data. The annual TFP growth (which picks up the exogenous factors that influence growth in an economy) was assumed to be 2 percent a year. This is somewhat conservative but not out of line with previous studies for India. The NCAER CGE model assumes TFP growth of 3 percent per year, as do the Energy and Research Institute (TERI) MoEF model and the IRADE AA model. However the same studies cite others that assume figures of between 1 and 3 percent. Given this range, an assumed value of 2 percent seems reasonable.

The assumed annual growth rate in real GDP from 2010 to 2030 is estimated at 6.7 percent. The economic growth (measured as an index) rises from 100 in 2010 to 367 in 2030. This is the conventional GDP growth scenario estimate (as per NCAER and recent IMF projections) without correcting for the implications of any new policy changes on pollution.

The standard GTAP model's structure has been modified to allow substitution between capital and energy (by increasing the elasticity of substitution from 0 to 0.5, as in the GTAP-E model). This modified version of the model is close to the energy version of the GTAP model (called GTAP-E) but does not include a nested structure in the energy block (which would require more data than were available).

Method for Estimating PM10 Emissions

The demands for different kinds of energy and the outputs of the different sectors were converted to PM10 emissions using corresponding emission coefficients ($\alpha_{i,j}$ and β_i, respectively).[18] The business-as-usual scenario generated PM10 emission estimates for 2010–2030 from fuel use and production activities as described in equation 5:

$$E = \sum_i \sum_j \alpha_{ij} C_{i,j} + \sum_i \beta_i XP_i \quad (5)$$

E = PM10 emissions
$C_{i,j}$ = demand for fuel products (j) in sector (i)
i = sector (firm, household, government)
j = energy products (coal, crude oil, refined oil and coal products, natural gas, and electricity)
$\alpha_{i,j}$ = emission coefficient associated with the consumption of one unit of energy product (j) by sector (i)
XP_i = production activity and process of sector (i)
β_i = emission coefficient associated with one unit of output in sector (i)

Both the consumer demand for energy products ($C_{i,j}$) and sectoral economic activity (XP_i) up to 2030 were estimated by the CGE model.

First, PM10 emission coefficients are taken from the Garbaccio, Ho, and Jorgenson (2000) study for China. This is presently the only source for these coefficients being mapped across sectors to the CGE model based on the GTAP database. The International Institute for Applied System Analysis (IIASA) reports PM10 and other secondary emissions for India, which corresponded to almost 8.7 million tons in 2010. The emission coefficients based on Garbaccio, Ho, and Jorgenson (2000) were updated to reproduce the aggregate PM10 and other small particles level for India in 2005 and then extrapolated following the growth assumptions in the business-as-usual scenario.

Table A4.3 in Appendix 4 represents the relative shares of PM10 emissions by sector and energy ($\alpha_{i,j}$).

Emissions from productive activities and the respective coefficients (β_i) were calculated as follows:

- First, the shares of production activities and process (XP) and energy use (C) related emissions in total emissions (E) were calculated as per the Garbaccio et al. (2000) study.

- Second, sector-specific emission coefficients in Garbaccio et al. (2000) were re-adjusted according to the GTAP classification in proportion to the sector's contribution to overall PM10 emissions and overall emission estimates from the IIASA model.

On the basis of equation (1) and the CGE simulations, the increase in PM10 emissions and other particulate emissions over time was calculated as a function of the demand for each type of energy by sector ($C_{i,j}$) and the economic expansion of production activities (XP_i).

Second, the emissions coefficients α_{ij} and β_i are modified over time to account for the improvements in the emission-capturing technologies through (1) a shift to cleaner coal (imported coal has lower emissions per unit of energy than domestic coal, and its share in the total amount of coal used in India is rising); and (2) other measures such as coal washing. These reductions in emissions are partly driven by administrative measures and partly by trade factors; such improvements are included in the business-as-usual scenario. The rates of decline in unit emission are for these reasons taken from microstudies (see Cropper et al., 2012). Further reductions in the coefficients may be achieved through a tax on PM10 and similar emissions. Such reductions in the coefficients reflect the impact of further pollution control measures that will be introduced as a result of the tax.[19]

The energy demand in value (US$) for four fuel types—coal, crude oil, oil and coal products, and natural gas—were obtained for 2010–2030 using the CGE model. This was converted into volume in terms of thousand tons of oil equivalent (TTOE) using appropriate factors.

4.5 Main Results

PM10 and Other Particle Emissions

Fossil fuel use, the primary cause of pollution, is expected to decrease under a business-as-usual scenario due to a declining share of coal in the overall energy demand (although coal would still dominate in 2030), improved emissions capture, and the shift to cleaner coal. Demand for refined oil products and electricity, however, will still increase considerably. As a result of rapidly growing economic activities, such as manufacturing and construction,

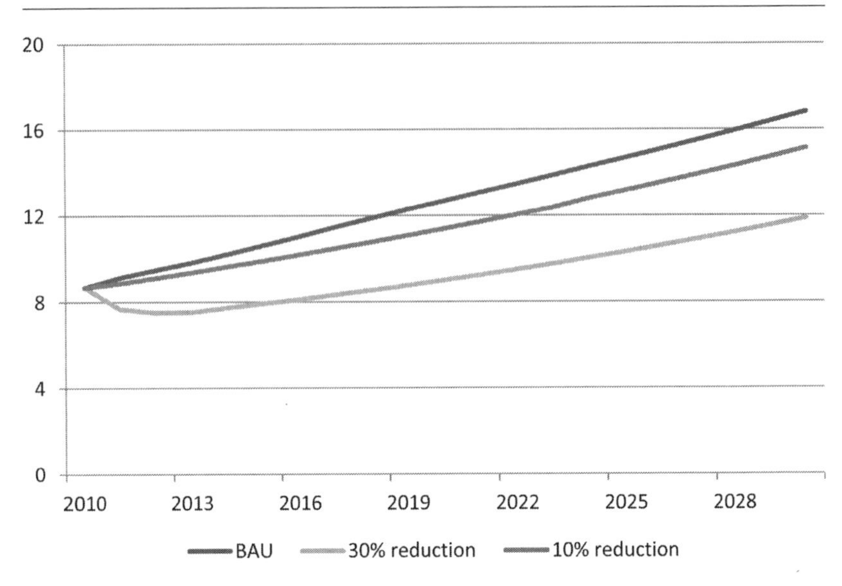

Figure 4.3 Total PM10 and similar emissions (business-as-usual [BAU] GDP Growth scenario) (in million tons)

and transportation, the share of emissions from productive activities in total PM10 emissions is expected to double by 2030. The total PM10 emissions under the business-as-usual scenario are estimated to increase from 8.7 million tons in 2010 to 16.8 million tons in 2030, an annual rate of increase of 1.9 percent against an annual GDP increase of 6.7 percent (Figure 4.3). Emissions will grow more slowly than GDP because of the exogenous factors noted above.

Conventional GDP Growth Scenario versus Green Growth Scenarios

Recall that the green growth scenarios seek to constrain particulate emissions through a menu of instruments that translate into 10 percent or 30 percent fewer emissions than under the business-as-usual scenario. They do this by imposing fuel or emission taxes, as already described, along with including other assumed reductions in emissions resulting from low-cost measures (especially in the green growth plus scenario) that are encouraged by tax policies that operate outside the scope of the model. The

Table 4.3 Comparing the business-as-usual scenario with a 10-percent and 30-percent reduction in PM

	2010	*2030*	*Percentage annual increase*	*GDP loss percent against Business-as-usual in 2030*	*Percent reduction in CO_2 against Business-as-usual in 2030*
Business-as-usual					
GDP in US$ billions, 2010	3,763	13,820	6.89		
CO_2 (million tons)	1,563	2,770	1.77		
PM10 (million tons)	8.68	16.81	1.94		
10-percent reduction via PM tax					20%
GDP in US$ billions, 2010	3,763	13,774	6.87	0.33	
CO_2 (million tons)	1,563	2,180	0.79		
PM10 (million tons)	8.68	15.12	1.03		
10-percent reduction via coal tax					10%
GDP in US$ billion, 2010	3,763	13,751	6.86	0.5	
CO_2 (million tons)	1,563	2,499	0.9		
PM10 (million tons)	8.68	15.24	1.03		
30-percent reduction via PM tax					60%
GDP in US$ billions, 2010	3,763	13,723	6.85	0.7	
CO_2 (million tons)	1,563	1,108	0.4		
PM10 (million tons)	8.68	11.86	1.02		
30-percent reduction via coal tax					30%
GDP in US$ billion, 2010	3,763	13,672	6.83	1.07	
CO_2 (million tons)	1,563	1,939	0.7		
PM10 (million tons)	8.68	11.84	1.02		

Note: Staff estimates

combined effect of the two drivers—the tax measures and other low-hanging-fruit measures—results in reduction in emissions of PM and CO_2. Table 4.3 shows the results, which are summarized as follows.

With the different tax regimes for a 10-percent particulate emission reduction, we have a lower GDP but the size of the reduction is modest. With a PM10 tax, GDP is about US$46 billion lower in 2030, representing a loss of 0.3 percent with respect to business as usual. The impact on GDP is greatest if we seek to achieve the PM target via a coal tax.

For a 30-percent particulate emission reduction, the GDP is about US$97 billion lower in 2030, representing a loss of 0.7 percent. This scenario suggests that even a substantial reduction in emissions can be achieved without compromising much on GDP and GDP growth rates if supported by adequate least-cost policy measures. Again the coal tax performs worse, with a GDP loss of 1.07 percent.

It should also be noted that the green growth plus scenario assumes, in addition to the taxes, some increase in investment for cleaner technologies. Such investments are associated with an increase in use of pollution-control techniques, modernization of the existing capital, and/or use of less-polluting capital over time with very low additional cost to the producer (see Box 4.3). These outside-the-model emission declines—which are assumed to be stimulated by the new investments and to have minimal economic costs—play a crucial role in the analysis of the environment–growth trade-offs for this scenario. They account for almost two-thirds of the PM10 reductions (20 out of 30 percent) in the green growth plus scenario. If we do not include these minimal-cost emissions savings from outside the model, there would be greater negative GDP impacts indicated by adjustments of inputs and outputs in the model. We would, however, argue that the stronger tax regime will result in enterprises looking to realize benefits from these low-cost mitigation measures.[20]

On the welfare side, health damages from PM are significantly reduced in the 30-percent reduction case when compared to a 10-percent reduction (Table 4.4). Savings range from US$24 billion from reduced health damages in the case of a 10-percent reduction (lower estimate) to US$105 billion in the case of a 30-percent reduction (upper estimate scenario). The central estimates are in the

Table 4.4 Health damage estimates for alternative scenarios

	2010	*2030*	*2030*	*2030*
Morbidity (US$ billion)		**Business-as-usual**	**10-percent PM reduction**	**30-percent PM reduction**
Lower	32.38	230.46	206.94	160.96
Central	46.12	328.37	294.84	229.28
Upper	72.39	515.24	462.64	359.83
Mortality (US$ billion)				
Lower	9.31	14.02	13.56	12.47
Central	14.87	22.36	21.63	19.90
Upper	20.39	30.65	29.65	27.29
Total (US$ billion)				
Lower	41.70	244.49	220.50	173.43
Central	60.99	350.73	316.48	249.18
Upper	92.78	545.90	492.29	387.11
Saving (US$ billion) from reduced health damages				
Lower			23.99	47.07
Central			34.25	67.30
Upper			53.60	105.18

Note: Staff estimates

US$34–67 billion range, which more or less offsets the GDP loss from the introduction of the tax.

The different tax regimes provide an important co-benefit in terms of substantial reduction in CO_2 emissions. We find the PM tax creates a bigger reduction in these emissions than the coal tax. Our calculations show that even with a value per ton of CO_2 of just US$10, the reduction in CO_2 for the 10-percent PM reduction case is worth US$59 billion, which is slightly more than the loss of GDP. For the 30-percent reduction case, the reduction in CO_2 is worth US$83 billion, slightly less than the loss of GDP.

Also, given our assumptions about the economy and environmental targets in 2030, the model gave us the percent of tax we have to apply on coal (first scenario) and coal and oil (second scenario) (see Table 4.2).

In terms of sector prices, we find that the energy-intensive sectors will be the most affected in 2030 under the various tax

Box 4.3 Technologies for Control of PM10 and CO_2 from Power Plants in India

Control of particulate emissions from power plants has been a concern in India for many years now, especially because of the high ash content of Indian coal, which is the primary fuel for the overwhelming majority of thermal power plants. Over the past two decades, various studies have been carried out to establish effective ways of dealing with these emissions over time (Lookman and Rubin, 1998; Kumar and Rao, 2003; Tata Energy Research Institute, 2003; Murty, Kumar, and Dhavala, 2006; Sengupta, 2007; Cropper et al., 2012).

Coal beneficiation is the process of removing the contaminants and the lower-grade coal to achieve a product quality that is suitable to the application of the end user—either as an energy source or as a chemical agent or feedstock.

A common term for this process is coal "washing" or "cleaning." According to Zamuda and Sharpe (2007), Indian coals are of poor quality and often contain 30–50 percent ash when shipped to power stations. In addition, over time the calorific value and the ash content of thermal coals in India have deteriorated as the better quality coal reserves have been depleted and surface mining and mechanization expanded. This poses significant challenges. Transporting large amounts of ash-forming minerals wastes energy and creates shortages of rail cars and port facilities. Coal washing reduces the ash content of coal, improves its heating value, and removes small amounts of other substances, such as sulfur and hazardous air pollutants. The benefits of using washed coal include, among others, reductions in particulate and sulfur emissions, reductions in flyash disposal costs, and reductions in the cost of transporting coal per unit of heat input. Use of washed coal may also reduce plant maintenance costs and increase plant availability.

Installing a washery for coal would entail an expenditure of around Rs 400 million for a 3 million tons per annum (MTPA) plant. According to Zamuda and Sharpe (2007) for a typical 500 megawatt (MW) plant, the use of washed coal with ash reduced from 38 percent to 30 percent could result in a 2-percent reduction in the cost of electricity generation, with savings averaging Rs 0.035 per kilowatt hour (kWh) of generated power, once various benefits to plant operation and reduced emissions are accounted for. Lookman and Rubin (1998) analyzed 174 plants across India and found that coal cleaning could result in savings in the range of US$75–150 million and US$15–25 million for existing plants by 2002 in terms of 1996 dollars. More recently, using updated figures from India's Central Electricity Authority for a particular plant in Rihand, Cropper et al. (2012) have estimated that the cost of electricity generation increases from Rs 1.206 to Rs 1.405 but did not take into account any of the other benefits that Zamuda and Sharpe quantified.

regimes. While the electricity, petroleum, chemical, and minerals sectors will be affected the most by a PM tax, metal products (e.g., iron and steel) will be most affected by a coal tax.

4.6 Conclusions

The study shows that policy interventions such as environmental taxes could potentially be used to yield positive net environmental benefits with minimal economic costs for India. The CGE analysis also shows that addressing "public bads" via selected policy instruments need not translate into large losses in GDP growth. The environmental cost model developed in this study can thus be used to evaluate the benefits of similar pollution-control policies and assist in designing and selecting appropriate targeted intervention policies (such as a SO_2 tax, a CO_2 tax, or emission trading schemes). Once the impact on ambient air quality of a policy to reduce particulate emissions is estimated, the tools used to calculate the health damages associated with particulate emissions can also be used to compute the welfare impacts of reducing them. The monetized value of the health benefits associated with each measure can be calculated using the techniques developed in this study and compared with the costs.

The comparisons made between the business-as-usual scenario and the green growth scenarios reveal that a low-carbon, resource-efficient greening of the economy should be possible at a very low cost in terms of GDP growth. This makes the green growth scenarios attractive compared to the business-as-usual scenario. A more aggressive low carbon strategy (green growth plus) comes at a slightly higher price tag for the economy while delivering higher benefits. The extent to which GDP growth would be affected under more severe cuts on polluting emissions can be determined by further study using the CGE model. On the other hand, the modest GDP impacts indicated in this study depend on the availability of minimal-cost mitigation options (energy efficiency improvements, embodied technological improvements, improved daily operating practices of boilers). With fewer such options, the GDP cost of hitting the 10-percent and 30-percent targets would be higher—potentially considerably higher. In evaluating the environment–growth trade-offs accordingly, a judgment must be made about the size and availability of such "low-hanging fruits" and appropriate incentives.

Both green growth scenarios have other important benefits. Most significantly they reduce CO_2 emissions, which have an important value. If we take that value at even a modest US$10 per ton, reflecting what might be gained in revenues from participation in emerging carbon abatement markets, India could realize an additional benefit of around US$59 billion with a PM10 tax. Global carbon models estimate that these emissions could be worth much more—US$50–120 per ton—by 2030. The green growth scenarios have other environmental benefits we have not included, especially in the areas of natural capital (elaborated in chapter 3). Finally the green growth scenarios may produce benefits for all—that is, they may have distributional advantages over the conventional scenario.[21]

The findings and conclusions of this study and the use of the CGE model should also be considered in the context of various assumptions/limitations:

- Only particulate emissions were analyzed; other local environmental issues were not considered.
- The baseline PM10 and other particulate emissions used in the CGE model were obtained from the IIASA literature on India and were not based on actual measurements.
- The CGE model did not separate health services from overall public services as an economic sector. The expected expansion of health services to address the increasing environmental health issues was not separately covered.[22]
- The CGE model has a medium- to long-term structure and therefore could not cover short-term fluctuations, such as oil-price volatility.

- Both production sectors (57) that cover agriculture, manufacturing, and services and households were represented as prototypes; thus the distributional environmental health impacts on different economic strata and geographic locations were not taken into account.

The study shows that the CGE model could be used as a tool for policy making. Being a general-equilibrium open-economy model, its strengths lie in the representation of intersectoral linkages both within and outside the country. At an economy-wide level, the CGE model makes it possible to determine whether growth objectives are compatible with the environmental objectives. The management of pollutants at the sectoral level can also be used to determine the abatement costs across the sectors. Distributional implications (winners versus losers) among the sectors could also be analyzed.

Further work using the CGE model after correcting for environmental health impacts would be useful in policy making. The present approach has the flexibility to incorporate multiple scenarios, such as the various scenarios in Parikh (2009), to determine the implications on GDP, which could be further corrected for environmental health impacts. The CGE approach described in this study was fairly detailed, with the 57 sectors tailored to India-specific parameters. The study recommends the use of this approach for the following possible scenarios:

- Including more energy sources so that it explicitly accounts for more renewable and nuclear energy.
- Considering higher levels of decarbonization and carbon capture and storage (CCS) as targets to be modeled and evaluated against the business-as-usual scenario.
- Examining different instruments (beyond the ones examined here) to achieve the shift from the business-as-usual scenario to an environmentally sustainable scenario.

Notes

1. CGE models are simulations that combine the abstract general equilibrium structure formalized by Arrow and Debreu (1954) with realistic economic data to solve numerically for the levels of supply, demand, and prices that support equilibrium across a specified set of markets. A CGE model consists of a set of equations representing the behavior of

all major sectors in an economy. These describe intersectoral linkages and the pattern of income and expenditure in the economy.

2. See Hertel et al. (1997) and https://www.gtap.agecon.purdue.edu/.
3. There are a number of studies that focus on estimating the greenhouse gas emissions trajectory of India for the next two decades, using a number of different modeling techniques including CGE modeling (Ojha, 2005, 2008; Ministry of Environment and Forests, 2007, 2009a; Planning Commission, Government of India, 2011). None of the models look at local pollutants such as PM10 and examine corresponding implications for economic growth and health. To that extent this study is first of its kind.
4. The standard version of the model represents the world economy in the form of 57 sectors/economic units trading with each other for 113 countries/regions. In this study, India is disaggregated from the rest of the regions and from the other South Asian countries.
5. See International Monetary Fund (2011).
6. See International Monetary Fund (2012), available at http://www.imf.org/external/pubs/ft/scr/2012/cr1296.pdf.
7. From the literature, the contribution to the costs of environmental degradation traditionally include not only PM10 and poor water supply and sanitation, but also groundwater depletion and soil degradation, which play a significant role in agriculture. These are not included in this study due to data and modeling constraints.
8. The cost of environmental degradation study that complements this study (Mani et al., 2012) finds that the health effects from particulate matter represent a loss of 1.7 percent of GDP, higher than any other type of environmental impact..
9. Recognizing the general uncertainty regarding the upper, central, and lower bound estimates are provided to indicate the ranges within which the actual health effects are likely to fall (Ostro, 1994). This is standard in environmental health literature.
10. The tax on PM10 also applies to secondary particles. Relatively generic coefficients are used to translate between fuel use and emissions, as distinct from more detailed and site-specific emissions coefficients; that is beyond the scope of the current model.
11. In this study we also conducted extensive research on cost and benefits of CO_2 mitigation and converted them to PM10 mitigation equivalents when needed. Our assumptions and results are aligned with the literature on critical parameters such as GDP elasticities of CO_2 mitigation, historical autonomous energy efficiency increase in India, and so forth.
12. The concept of "uniform rollback" states that the percentage change in pollutant emissions can be assumed to be equal to the percentage change in pollutant concentration. This assumption invariably involves a simplification of how emissions affect air quality; how much of a simplification depends on specific circumstances.
13. Autonomous energy efficiency (kilograms CO_2 emitted per unit of GDP in 2000 dollars) improved by 1 percent per year between 1980

and 2008 (World Development Indicators World Bank, 2011, and our business-as-usual scenario reproduces the same trend.

14. New substitution elasticity between capital and energy was introduced into the standard GTAP model to capture this effect. This is based on the notion that technical progress is entirely embodied in the design and operating characteristics of new capital plant and equipment. For example, the energy saving effects of embodied technical progress depends critically on the rate at which new investment goods diffuse into the economy. By introducing substitution between capital and energy in the model, we mitigate CO_2 emissions by 20 percent. India would have emitted 3,246 metric tons in 2030, but with the substitution emits only 2,631 metric tons under the business-as-usual scenario).
15. Although most countries use technical standards to curb air pollutants, modeling the effect of market-based instruments is useful because they favor allocation through relative prices. This is consistent with India's recent approach to use market based instruments to deal with air pollution. On July 1, 2010, the government of India introduced a nationwide coal tax of 50 rupees per metric ton (US$1.07 per ton) of coal both produced and imported into India. The tax raised Rs 25 billion (US$535 million) for the financial year 2010–2011. Many consider this coal tax is a step toward helping India meet its voluntary target to reduce the amount of carbon dioxide released per unit of GDP by 25 percent from 2005 levels by 2020. Further, India's federal cabinet on April 12, 2012, approved a proposal to change the method used to calculate the royalty that coal miners pay to state governments, imposing a flat 14-percent tax based on prices.
16. Environmental taxes are corrective measures for dealing with the environmental "externality" first studied by Pigou (1932). A Pigouvian approach sets taxes equal to the marginal damage caused to the environment by the production process thereby "internalizing" the full social marginal costs.
17. A more formal representation of this can be found in Conrad and Henseler-Unger (1986).
18. Emission coefficients vary through time to reflect technological change, modernization of power plants, improved energy efficiency, and India's emission abatement levels (on the basis of 1 percent annual increase on average in business-as-usual reported in WDI statistics).
19. The PM10/CO_2 elasticity varies across scenarios; the average is 1.62 which means that 1 unit of CO_2 abatement will bring 1.62 units of PM10 abatement.
20. As a result of tax policies, private firms are expected to invest in clean technologies, either financed by foreign direct investment (FDI) or through domestic investments. This investment may even generate new activity sectors if environment-friendly technologies are domestically produced. According to the model estimations these new investments will generate a value added equivalent of 0.8–1.2 percent of GDP in different scenarios that we simulated.

21. Improving air quality is a public good. Even if poor air quality affects all equally, an improvement has a bigger proportional benefit to the poor, and there is evidence that the poor are more affected by air pollution.
22. Health services are in the same category as education and defense: public services. They are separated from other services provided by the private sector, such as trade, transport, and so forth.

Chapter 5

Way Forward: Striving for Green Growth

Although the past decade of rapid economic growth has brought many benefits to India, the environment has suffered. Moreover, the large scale and rapid growth of the Indian economy have exposed a growing population to serious air and water pollution (though in some cases pollution levels have fallen). Without stringent pollution control the environmental damage is likely to worsen. While the overall policy focus should be on meeting basic needs and expanding opportunities for growth, these goals should not come at the expense of unsustainable environmental degradation. Green growth is growth that is environmentally sustainable, uses natural resources efficiently, and minimizes pollution and environmental impacts.

The findings of this study suggest that environmental performance does not automatically improve with national income. Policies will be required to prevent and remedy growth obstacles and negative impacts on welfare from unsustainable practices. Otherwise it may be impossible or prohibitively expensive to clean up if we were to wait for the country to achieve a suitable level of prosperity.

The cost of degradation exercise undertaken here could be instrumental in moving the environmental debate beyond the ministries of environment to reach other sectoral ministries, especially the finance ministry. Over the past decades, COED analyses like this have had major impacts on decision makers in a number of countries in terms of influencing national policy dialogue, increasing environmental investments, and strengthening the capacity of national institutions in environmental valuation. COED analysis also highlights the need to incorporate the results of the environmental valuation into decision making at the sectoral and national levels, so that the environmental costs and benefits are mainstreamed into national and local planning processes.

Typically, conventional measures of growth do not adequately capture the environmental costs, which have been found to be particularly severe at the current rapid growth rates. Therefore, it is imperative to calculate green gross domestic product (green GDP) as an index of economic growth with the environmental consequences factored in. The government of India has already set a target year of 2015 to release green GDP with a view to bring environmental concerns into mainstream growth accounting. The last two decades witnessed measures by India to take care of the fiscal sustainability of its growth. The next focus should be to ensure the ecological sustainability of its growth. This commitment is a landmark one and augurs well for an environmentally sustainable future.

A scientific committee chaired by Sir Partha Dasgupta has recently been given the mandate by the government of India to develop a framework for green national accounts, identify data gaps, and prepare a road map to implement the framework. A report developed by the committee will advise on methods to reformulate and re-create national accounts and provide guidelines for preparing them. When released, the green GDP data will have a number of methodological challenges and will require constant refinement. This is only to be expected. But deliberating on the green GDP data on an annual basis will drive better policy and decision making. It will also ensure better control in the use of natural resources and reduce the environmental deterioration that appears concomitant to economic progress. The entire process that is used to generate the GDP should be reviewed through a green lens. In particular, the ministries involved in using the natural resources, such as mines and minerals and forests, as well as the sectors contributing to the environmental degradation, such as power, industry, and urban development, should be actively engaged in this green GDP exercise in order to be cognizant of environmental costs.

For an environmentally sustainable future, India also needs to value its natural resources and ecosystem services to better inform policy and decision making. Given that India is a hotspot of unique biodiversity and ecosystems, it is necessary to have a structured approach to such valuation in the context of rapid growth. These valuations are yet to be systematized and, more importantly, used in policy making. Such a valuation effort will lead to better decision making with regard to other development initiatives that could have a negative impact on India's biodiversity and ecosystems.

The present study draws attention to the global economic benefits of biodiversity to highlight the growing costs of biodiversity loss and ecosystem degradation and brings together expertise from the fields of science, economics, and policy to enable practical actions in moving forward. In this regard, research on valuation of ecosystem services should be updated periodically, and more importantly, the outcomes of such research should be used in defining policy. Instruments such as payment for environmental services (PES) to generate revenue for biodiversity and ecosystem conservation should be introduced. The Thirteenth Finance Commission acknowledged some of these issues, but its recommendations, such as grants for states to preserve forests, are inadequate to deal with the size and complexity of the problem. Admittedly, the commission was hamstrung by the lack of a database that provides accurate and timely information on India's natural wealth at the national and subnational level.

The study also shows interventions to improve the environment in India are likely to yield positive net benefits. Indeed, one of the advantages of the CGE modeling developed in this study is that it can be used to evaluate the benefits of specific pollution-control policies and assist in designing and selecting appropriate targeted intervention policies. Once the impact on ambient air quality of a policy to reduce particulate emissions has been calculated, the tools used to calculate the health damages associated with particulate emissions can be used to determine the benefits of reducing them.

CGE and other similar modeling thus could be gainfully used as a tool for assessing trade-offs between various green policy options. As a general-equilibrium open-economy model, CGE's strengths also lie in the representation of intersectoral linkages both within and outside the country. At an economy-wide level, CGE modeling makes it possible to determine whether growth objectives are compatible with the environmental objectives. The management of pollutants at the sectoral level can also be used to determine the abatement costs across the sectors. Distributional implications (winners versus losers) among the sectors could also be analyzed.

This study shows that policy interventions such as environmental taxes could potentially be used to yield positive net environmental benefits with minimal economic costs for India. The government of India has already introduced a pilot emission trading scheme for particulate emissions in three states. This indicates a positive move toward embracing market-based instruments for emission reduction. The results from the pilot could be extremely valuable in scaling up such a system across the country in various sectors.

Past policies and decisions have been made in the absence of concrete knowledge of the environmental impacts and costs. By providing new, quantitative information based on research under Indian conditions, this study has aimed to reduce this information gap. At the same time, it has pointed out that substantially more information is needed in order to understand the health and nonhealth consequences of pollution. It is critically important that existing air quality, health, and environmental data be made publicly available so the fullest use can be made of them. This would facilitate conducting periodic studies on the impacts of air pollution on human health. Furthermore, surveillance capacity at the local and national levels needs to be expanded to improve the collection of environmental data, especially data on various aspects of air quality. These efforts will further improve the analysis begun in this study.

To support the move toward introducing economic instruments such as green accounting, ecosystem valuation, emissions trading, and eco-taxes and even assessing impact of climate change mitigation and adaptation strategies, it would be worthwhile to consider strengthening the scope and usage of environmental economics in policy making. Institutionalizing this would further ensure that we have a correct understanding of the costs of environmental degradation and are able to assess the costs and benefits of various programs and policies, as well as the trade-offs of alternative growth scenarios. This could be along the lines of the U.S. Environmental Protection Agency's National Center for Environmental Economics (NCEE), which offers a centralized source of technical expertise to the EPA, as well as other federal agencies, Congress, universities, and organizations. NCEE's staff specializes in analyzing the economic and health impacts of environmental regulations and policies, and assists the EPA by informing important policy decisions with sound economics and other sciences. NCEE also contributes to and manages the EPA's research on environmental economics to improve the methods and data available for policy analysis.

India has a structured institutional framework for environmental management. This framework is unable to cope with the present environmental challenges and needs considerable strengthening in the context of continued and rapid economic growth. In addition, given the growing population, increased urbanization, and drive toward greater economic well-being, there is little or no doubt that the environmental challenges will continue to increase steadily. That being the case, there is no doubt that substantive strengthening of the policy and institutional framework will be required in order to manage the greater challenges.

Green growth strategies are needed to break the pattern of environmental degradation and natural resource depletion that are too often the consequence of economic growth and to avoid locking the economy into unsustainable patterns. This is required across the board—that is, with policies and instruments, environmental organizations, human resources and capacity building, and financing and investments. It is important to recognize that the effectiveness of the institutional framework does not lie solely with the environmental sector alone. As the nature of the subject is cross-cutting, sectors that cause environmental impacts and those vulnerable to environmental degradation are also relevant to achieve effective environmental management. Also, the purpose of the institutional framework is to serve its people and society with better environmental quality. There are four key take-aways from the study.

Green growth is necessary: At the current rate of degradation, environmental sustainability could become the next major challenge as India surges along its projected growth trajectory.

Green growth is affordable: A low-emission, resource-efficient greening of the economy should be possible at a very low cost in terms of GDP growth. It also promises to deliver greater co-benefits.

Green growth is desirable: For an environmentally sustainable future, India needs to value its natural resources and ecosystem services to better inform policy and decision making especially since India is a hotspot of unique biodiversity and ecosystems.

Green growth is measurable: Conventional measures of growth do not adequately capture the environmental costs as well as benefits of ecosystem services. Therefore, it is imperative to calculate green Gross Domestic Product (green GDP) as an index of economic growth with the environmental consequences factored in.

Appendix I

Methodology of Environmental Health Losses Valuation

A1.1 Outdoor Air Pollution

Mortality

Based on the current status of worldwide research, the risk ratios or concentration-response coefficients from Pope et al. (2002) were considered likely to be the best available evidence of the mortality effects of ambient particulate pollution (PM2.5). These coefficients were applied by the WHO in the *World Health Report 2002*, which provided a global estimate of the health effects of environmental risk factors. Pope et al. (2002) provide the most comprehensive and detailed research study to date on the relationship between air pollution and mortality. That study confirms and strengthens the evidence of the long-term mortality effects of particulate pollution found by Pope et al. (1995) and Dockery et al. (1993). The study found a statistically significant relationship between levels of PM2.5 and mortality rates, controlling for all the factors.

Damages due to anthropogenic factors are measured from a baseline PM2.5 concentration, which we set equal to 7.5 $\mu g/m^3$ (as in WHO, 2002a). This is considered to be the level one would find in the natural environment.

Morbidity

While the mortality effects are based on PM2.5, the morbidity effects assessed in most worldwide studies are based on PM10. Concentration-response coefficients from Ostro (1994), Ostro and Chestnut (1998) and Abbey et al. (1995) have been applied to estimate these effects. Ostro (1994) reviewed worldwide studies and based on that estimated concentration-response coefficient for restricted activity days (Ostro and Chestnut, 1998), and Abbey et al. (1995) provided estimates of chronic bronchitis associated with particulates (PM10). The mortality and morbidity coefficients are presented in Table A1.1 based on these estimates.

Baseline concentration for the application of the concentration-response functions was set at 7.5 μg/m^3 for PM2.5 (as for mortality).

Table A1.1 Urban air pollution concentration-response coefficients

Annual health effect	*Concentration-response coefficient*	*Per 1 μg/m^3 annual average ambient concentration of*
Long-term mortality (change in cardiopulmonary and lung cancer mortality)	0.8% *	PM2.5
Acute mortality children under five (change in ARI deaths)	0.166%	PM10
Chronic bronchitis (change in annual incidence)	0.9%	PM10
Respiratory hospital admissions (per 100,000 population)	1.2	PM10
Emergency room visits (per 100,000 population)	24	PM10
Restricted activity days (change in annual incidence)	0.475%	PM10
Lower respiratory illness in children (per 100,000 children)	169	PM10
Respiratory symptoms (per 100,000 adults)	18,300	PM10

* Mid-range coefficient from Pope et al. (2002) reflecting a linear function of relative risk. In the analysis, however, we used a log-linear.

Sources: Pope et al. (2002) and Ostro (2004) for the mortality coefficients; Ostro (1994) and Ostro and Chestnut (1998) and Abbey et al. (1995) for the morbidity coefficients.

As in Ostro (1994) there is no threshold for morbidity, estimated utilizing PM10 concentrations.

Expressing Health Effects in DALYS

The health effects of air pollution can be converted to disability-adjusted life years (DALYs) to facilitate a comparison with health effects from other environmental risk factors.

Estimation of Urban Population and of Annual Average PM10 Concentrations

The last available census was in the year 2001, since it is conducted once every decade. The population figures for India are slightly outdated given the rapid growth in population and urban areas around the country. Consequently this study uses population figures from the 2001 census that have been projected to 2009 using UN Population Fund projections for urban areas in India. The UN database provides annual population growth rates for selected cities, which have been used for the population projections. For cities without these growth rates, we take the average annual growth rate and project the population to 2009.

In this study we focus only on cities with a population of 100,000 and greater. Since the baseline population is from the 2001 census, there are many cities that have achieved a population of 100,000 since 2001 and have not been included in the study. This can be updated once the figures for the latest census are released in 2011.

Pollution data for all cities, wherever available, was taken from the Central Pollution Control Board's (CPCB) Environmental Data Bank website for the year 2008. Health damage estimates for PM10 were calculated based on observations for the year 2008. The average concentration was calculated by taking the arithmetic average for all available observations in the year 2008 (Table A1.2). The local state pollution control board is in charge of measuring pollution levels in each city at each of the monitoring stations. There are supposed to be 104 observations for each monitoring station in each city annually, which is roughly two readings a week at each monitoring station. The frequency of observations depends on the pollution control board officials at the city level. Once the data has been collected, it is loaded on the CPCB Environmental Data Bank website by the local officials.

Table A1.2 Average annual concentrations of PM10 (μg/m^3) and population for major Indian cities

	2008 PM10 concentration projections (μg/m^3)	*2009 population projections (thousands)*
Meerut	313	1,413
Yamunanagar	301	340
Ludhiana	271	1,668
Ghaziabad	236	791
Firozabad	222	418
New Delhi	214	21,331
Delhi	214	13,010
Kanpur	210	3,195
Indore	196	2,093
Raipur	192	906
Lucknow	189	2,723
Amritsar	189	1,252
Satna	188	248
Agra	188	1,643
Allahabad	181	1,238
Ranchi	175	1,078
Jamshedpur	172	1,341
Chandrapur	170	349
Guwahati	164	1,015
Faridabad	163	955
Gwalior	162	1,008
Jalandhar	150	884
Jodhpur	148	1,026
Noida	148	226
Alwar	144	325
Jabalpur	136	1,324
Asansol	135	1,372
Durgapur	133	658
Dhanbad	131	1,285
Jhansi	130	569
Nagpur	128	2,526
Bombay/Mumbai	127	19,460
Jaipur	126	3,012
Kota	125	837
Patna	120	2,231
Nellore	118	489
Sagar	115	397
Hisar	112	280
Bhilai Nagar	109	1,059
Dehradun	109	569
Korba	107	355

	2008 PM10 concentration projections (μg/m³)	*2009 population projections (thousands)*
Varanasi	106	1,391
Rajkot	105	1,304
Hubli-Dharwad	103	918
Calcutta/Kolkata	103	17,032
Bhopal	102	1,780
Raurkela	102	616
Ujjain	101	560
Bangalore	97	6,982
Vijayawada	96	1,171
Chandigarh	95	1,012
Jamnagar	95	590
Pune	94	3,854
Udaipur	91	477
Ahmedabad	88	5,531
Surat	88	3,982
Ramagundam	87	331
Bhubaneswar	86	875
Kolhapur	84	647
Imphal	84	313
Hyderabad	84	6,551
Dewas	84	254
Cuttack	81	680
Visakhapatnam	80	1,575
Nashik	79	1,524
Solapur	79	1,092
Salem	78	901
Vadodara	77	1,810
Shillong	72	345
Gulbarga	71	480
Kurnool	71	426
Coimbatore	71	1,748
Warangal	69	723
Amravati	66	651
Baleshwar	66	157
Thiruvananthapuram	64	981
Madras/Chennai	63	7,347
Haldia	61	155
Mangalore	60	659
Thane	58	1,241
Dibrugarh	56	194
Sangli	55	562
Shimla	54	171

(Continued)

Table A1.2 (Continued)

	2008 PM10 concentration projections (μg/m³)	*2009 population projections (thousands)*
Pondicherry	50	620
Sambalpur	50	299
Hassan	50	168
Mysore	49	914
Kakinada	48	506
Kottayam	46	257
Kochi	43	1,538
Madurai	42	1,311
Aizawl	36	240
Tirupati	34	292
Kozhikode	34	966
Belgaum	33	622
Palakkad	30	278

Source: Staff estimates.

The PM10 concentrations data from CPCB are the best available because this is the monitoring agency with the most widespread network. The CPCB, however, does not cover all of the cities on our list. For cities not monitored by CPCB, the PM10 concentrations have been estimated for 2009 (Table A1.3).

Table A1.3 PM10 estimates and population for cities with no monitoring data

	2008 PM10 concentration projections (μg/m³)	*2009 population projections (thousands)*
Cities with population above 1 million		
Haora	154	1,469
Durg	122	1,132
Aurangabad	118	1,152
Kalyan	45	1,568
Srinagar	20	1,176
Cities with population between 500,000 and 1 million		
Bareilly	164	842
Gorakhpur	158	781
Saharanpur	150	579
Bikaner	149	643
Ajmer	142	622
Moradabad	120	814

	2008 PM10 concentration projections ($\mu g/m^3$)	*2009 population projections (thousands)*
Bhavnagar	105	626
Bokaro Steel City	102	616
Akola	93	507
Guntur	81	728
Ulhasnagar	70	570
Tiruchchirappalli	63	980
Aligarh	62	832
Tiruppur	59	759
Erode	59	559
Bhiwandi	58	824
Tirunelveli	58	567
Kannur	55	717
Malegaon	54	529
Rajahmundry	53	620
Kollam	39	560
Cities with population between 100,000 and 500,000		
Muzaffarpur	283	373
Darbhanga	217	337
Panipat	174	295
Shahjahanpur	169	402
Karnal	169	272
Baranagar	162	347
Arrah	161	243
Panihati	156	426
Muzaffarnagar	156	383
Rampur	156	377
Wadhwan	156	257
Sirsa	156	174
Kamarhati	155	412
Bhiwani	155	188
Patiala	154	392
Maunath Bhanjan	152	211
Hugli-Chinsura	151	235
Bahraich	151	209
Naihati	151	205
Unnao	151	166
Bathinda	149	246
Bhagalpur	148	402
South Dum Dum	147	360
Rae Bareli	147	201
Shivapuri	147	167

(Continued)

Table A1.3 (Continued)

	2008 PM10 concentration projections (μg/m³)	*2009 population projections (thousands)*
Bally	146	285
Chandan Nagar	146	186
Faizabad	145	273
Ambala	145	216
Serampore	145	212
Budaun	145	180
Titagarh	145	176
Gurgaon	144	210
Baharampur	144	195
Gaya	142	455
Bharatpur	142	242
Ganganagar	141	250
Sonipat	141	222
Pilibhit	141	165
Hapur	140	226
Modinagar	139	191
Amroha	137	212
Etawah	137	192
Guna	137	155
Moga	136	171
Jaunpur	135	210
Alipurduar	134	159
Sikar	133	229
Purnia	133	212
Bidar	132	205
Beawar	131	165
Fatehpur	130	182
Bhilwara	129	284
Katihar	128	239
Barrackpur	128	206
Siliguri	127	335
Morena	127	227
Hathras	126	175
Kanchrapara	126	155
Chapra	125	212
Hardwar	123	290
Sitapur	121	188
Yavatmal	118	188
Anand	117	270
Murwara (Katni)	117	253
Mahesana	117	170
English Bazar	115	274

	2008 PM10 concentration projections (μg/m³)	*2009 population projections (thousands)*
Bharuch	115	215
Nadiad	113	263
Dabgram	112	227
Mathura	110	365
Hoshiarpur	110	190
Bhind	110	170
Pathankot	109	198
Tonk	107	155
Pali	106	211
Rewa	106	199
Bulandshahr	106	197
Wardha	104	159
Damoh	103	162
Sambhal	102	233
Abohar	102	166
Ahmednagar	101	343
Haldwani-cum-Kathgodam	101	161
Batala	101	160
Bhatpara	99	471
Mirzapur-cum-Vindhyachal	99	262
Raiganj	99	246
Bid	98	174
Brahmapur	97	325
Gondiya	96	169
Nanded	95	478
Bellary	95	379
Nizamabad	95	372
Porbandar	95	248
Agartala	95	243
Gandhinagar	94	191
Santipur	94	170
Kharagpur	93	409
Ondal	93	327
Proddatur	92	207
Bilaspur	91	355
Ratlam	91	303
Raniganj	91	241
Balurghat	91	195
Krishnanagar	89	187
Guntakal	89	166

(Continued)

Table A1.3 (Continued)

	2008 PM10 concentration projections (μg/m³)	*2009 population projections (thousands)*
Jorhat	88	173
Cuddapah	87	334
Adoni	86	210
Kamptee	86	196
Ranaghat	86	196
Khandwa	85	224
Godhra	85	156
Bihar Sharif	84	311
Nabadwip	84	241
Medinipur	81	194
Farrukhabad Cum Fategarh	80	323
Munger	80	232
Rohtak	79	334
Parbhani	79	294
Phusro	79	220
Jalna	78	270
Karimnagar	78	230
Nandyal	78	185
Burhanpur	77	267
Patratu	77	170
Raichur	76	264
Silchar	76	178
Rajnandgaon	75	194
Barddhaman	74	379
Puri	74	193
Hospet	71	208
Bhusawal	70	247
Kolar Gold Fields	69	242
Kothagudem	69	158
Tenali	68	222
Mahbubnagar	68	181
Khammam	67	230
Latur	66	305
Chiral	66	221
Ongole	66	199
Bankura	66	178
Valsad	66	173
Tiruvannamalai	66	169
Vellore	65	480
Habra	64	304
Machilipatnam	64	246

	2008 PM10 concentration projections ($\mu g/m^3$)	2009 population projections (thousands)
Dhule	63	430
Jalgaon	63	374
Arcot	62	177
Ichalakaranji	60	365
Thanjavur	60	312
Thalassery	60	160
Kanchipuram	59	264
Basirhat	59	157
Tumkur	58	278
Chittoor	58	206
Patan	58	186
Morvi	58	186
Kumbakonam	57	233
Neyveli	57	196
Bhimavaram	57	187
Eluru	56	329
Hindupur	56	162
Cuddalore	55	223
Gudivada	55	157
Junagadh	53	258
Mandya	52	186
Chitradurga	52	160
Vizianagarm	51	274
Bhuj	51	187
Karur	51	176
Karaikkudi	51	171
Davangere	50	444
Bijapur	50	298
Dindigul	50	282
Gandhidham	50	162
Allappuzha	48	409
Bhadravati	48	231
Tuticorin	47	433
Shimoga	46	298
Rajapalaiyam	46	176
Thrissur	45	425
Sivakasi	45	158
Navsari	44	295
Nagercoil	44	294
Anantapur	44	270
Cherthala	42	205
Udupi	38	182

(Continued)

Table A1.3 (Continued)

	2008 PM10 concentration projections (μg/m³)	*2009 population projections (thousands)*
Pollachi	37	196
Gadag-Betgeri	36	207
Guruvayur	36	183
Vadakara	34	158
Kanhangad	33	183
Malappuram	31	220
Valparai	16	165

Source: PM estimated by Aarsi Sagar based on the World Bank Model.II.

The monitoring stations in cities are placed in three areas: residential, industrial, and sensitive. Residential areas are locations with housing; industrial areas are locations with mostly industries; and sensitive areas are locations either with monuments or with biodiversity and zoo parks. Depending on the city and the prominent activities, some have a greater number of monitoring stations in residential areas, and other have a majority of the monitoring stations in industrial areas. Cities with monuments will have monitoring stations in these sensitive areas. Therefore, the distribution of these monitoring stations by type is not constant in each city.

Population in each city was estimated from 2001 Census data.[1]

Estimating Mortality and Morbidity Effects of PM

Concentration Response Functions

As noted, PM is the main form of outdoor air pollutant with health impacts. Pope et al. (2002) estimated relative risk for the linear function for cardiopulmonary mortality. That is

$$RR = \exp(\beta(X - X_0)) \tag{1}$$

Where *RR* stands for relative risk for cardiopulmonary mortality, X is the observed PM2.5 concentration and X_0 is a background PM2.5 concentration, which we set equal to 7.5 μg/m³ (as in WHO, 2002a). Based on that, one finds the increase to be 6–9 percent in cardiopulmonary mortality, and 8–14 percent for lung cancer per 10 μg/m³ of PM2.5. The mortality coefficient in Table A1.1 is a combination of the cardiopulmonary and lung cancer mortality risk ratios.

However, for higher PM2.5 concentrations than Pope et al. (2002) considered in their analysis and such as those found in India, Ostro (2004) proposed to use log-linear relative risk function from cardiopulmonary mortality reflecting the uncertainty about the health impact with higher PM2.5 concentration. The log-linear relative risk function for cardiopulmonary mortality has the form:

$$RR = [(1 + X)/(1 + X_0)]^{\beta}, \quad (2)$$

where β is equal to 0.15515 (Ostro, 2004). In order not to overestimate the impacts of higher concentrations, we used the log-linear form. The difference between the two can be seen in Figure A1.1.

The share of cardiopulmonary and lung cancer deaths in total mortality varies, sometimes substantially, across countries. It may therefore reasonably be expected that the risk ratios for cardiopulmonary and lung cancer mortality provide more reliable estimates of mortality from PM2.5 than the risk ratio for all-cause mortality when the risk ratios are applied to countries other than the United

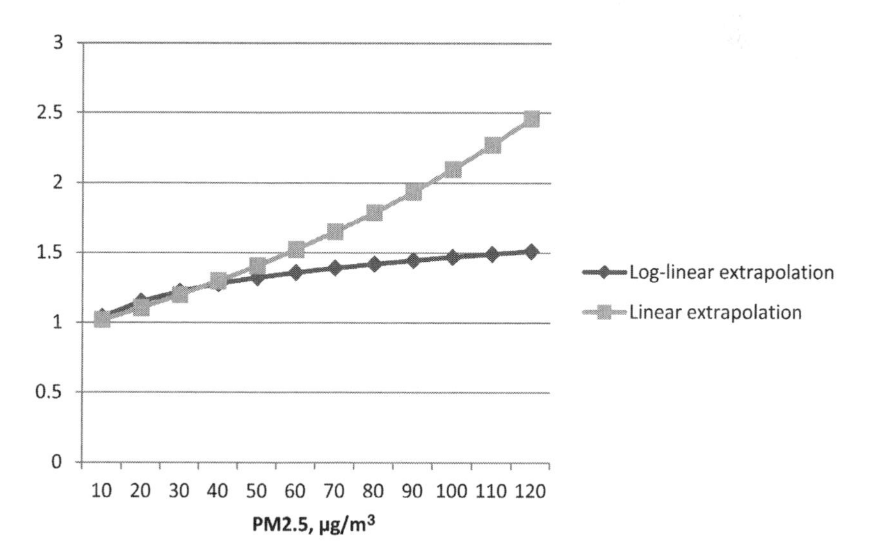

Figure A1.1 Alternative concentration-response curves for mortality from cardiopulmonary diseases

States. The cardiopulmonary risk ratio is therefore used in this report.

In order to apply the mortality coefficients in Table A1.1 to estimate mortality from urban air pollution, baseline data on total annual cardiopulmonary and lung cancer deaths are required. For this purpose, we applied the Office of Registrar General and Commissioner of Census (2004) report on causes of death in India. An urban crude mortality rate of 5.9 per 1,000 was applied, along with an average cardiopulmonary and lung cancer mortality rate of 35 percent of total deaths. An annual rate for ARI deaths for children under five of 22 percent was applied per the report of the Office of Registrar General and Commissioner of Census (2004).

Estimating Morbidity Cases and Costs of PM10

A number of issues need to be addressed with respect to the morbidity costs.

1. *Baseline Incidence.* To apply the coefficients in Table A1.1, we need information in some cases on the baseline rates of the incidence of the health item. This information is not available for chronic bronchitis in India so the rate was taken from WHO (2001) and Shibuya, Mathers, and Lopez (2001) for the Sear D regions of WHO.[2] Since this figure is taken from outside India and applied nationally, it has not been possible to provide city-specific incidence rates of chronic bronchitis.
2. *Restricted Activity Days.* In the case of restricted activity days, the background values were estimated from ARI prevalence in the adult population (see chapter 2, section 5 on health loss from indoor air pollution). From international experience each illness was estimated to last seven days, of which five were counted as restricted activity days.
3. *Other morbidity health endpoints.* These are hospital admissions of patients with respiratory problems, emergency room visits (or hospital outpatient visits), lower respiratory infections in children, and respiratory symptoms. These are the most common health endpoints considered in worldwide studies on air pollution. The coefficients are expressed as cases per 100,000 in the absence of incidence data for India. It should be noted that it would be preferable to have incidence data and use coefficients that reflect percentage change in incidence.

Increases in asthma attacks among asthmatics have also been related to air pollution in many studies. This calculation, however, requires data on the percentage of the population that are asthmatic and frequency of asthma attacks, which is not readily available for India.

4. *Use of DALYS and calculation of DALYs per health effect.* The base case numbers of DALYs per endpoint given in Table A1.4 are based on the disability weights and average duration of each illness. The weights for lower respiratory illness (LRI) and chronic bronchitis (CB) are presented by the US National Institutes of Health.[3] Disability weights for the other morbidity endpoints are not readily available; Larsen (2004a) provides estimates based on weights for other comparable illnesses.[4] Average duration of CB is estimated based on age distribution in India and age-specific CB incidence in Shibuya, Mathers, and Lopez (2001). Years lost to premature mortality from air pollution is estimated from age-specific mortality data for cardiopulmonary and lung cancer deaths, and have been discounted at 3 percent per year. Average duration of illness for the other health endpoints is from Larsen (2004a). The details are summarized in Table A1.4.

Table A1.4 Calculation of DALYs per case of health effects for outdoor air pollution

Health effects	*Disability weight*	*Average duration of illness*
Mortality	1.0	7.5 years lost or 70 years lost for children under five
Lower respiratory illness (children)	0.28	10 days
Respiratory symptoms (adults)	0.05	0.5 days
Restricted activity days (adults)	0.1	1 day
Emergency room visits	0.30	5 days
Hospital admissions	0.40	14 days*
Chronic bronchitis	0.2	20 years

* Includes days of hospitalization and recovery period after hospitalization.

5. *Baseline data to value costs per case of illness.* These data are summarized in Table A1.5. Some of these data require explanation. The value of time for adults is based on urban wages. Economists commonly apply a range of 50–100 percent of wage rates to reflect the value of time. The rate of Rs 200 per day is an average urban wage in India. It was estimated using the India 2011 data on household monthly income from wages.[5] Seventy-five percent of this rate has been applied for both income-earning and nonincome-earning individuals. There are two reasons for applying the rate to nonincome-earning individuals. First, most nonincome-earning adult individuals provide a household function that has a value. Second, there is an opportunity cost to the time of nonincome-earning individuals, because they could choose to join the paid labor force.[6]

Table A1.5 Baseline data for cost estimation: Outdoor air pollution

	Baseline	*Source*
Cost data for all health endpoints:		
Cost of hospitalization (Rs per day)	980	NSS (2004), and per consultations with medical service providers and health authorities
Cost of emergency visit (Rs): urban	800	
Cost of doctor visit (Rs) (mainly private doctors): urban	800	
Value of time lost to illness (Rs per day)	150	75 percent of urban wages in India
Chronic bronchitis (CB):		
Average duration of illness (years)	20	Based on Shibuya et al. (2001)
Percent of CB patients hospitalized per year	1.5%	From Schulman et al. (2001) and Niederman et al. (1999)
Average length of hospitalization (days)	10	
Average number of doctor visits per CB patient per year	1	
Percent of CB patients with an emergency doctor/hospital outpatient visit per year	15%	

	Baseline	Source
Estimated lost work days (including household work days) per year per CB patient	2.6	Estimated based on frequency of doctor visits, emergency visits and hospitalization
Annual real increases in economic cost of health services and value of time (real wages)	2%	Estimate
Annual discount rate	3%	Applied by WHO for health effects
Hospital admissions:		
Average length of hospitalization (days)	6	Estimates
Average number of days lost to illness (after hospitalization)	4	
Emergency room visits:		
Average number of days lost to illness	2	
Restricted activity days:		
Average number of days of illness (per 10 cases)	2.5	
Lower respiratory illness in children:		
Number of doctor visits	1	
Total time of care giving by adult (days)	1	Estimated at 1–2 hours per day

A1.2 Impacts from Inadequate Water, Sewage, and Hygiene

Background

Inadequate quantity and quality of potable water supply, sanitation facilities and practices, and hygiene conditions are associated with various illnesses both in adults and children. Esrey et al. (1991) provides a comprehensive review of studies documenting this relationship for diseases such as schistosomiasis (bilharzia), intestinal worms, diarrhea, and so forth. Fewtrell and Colford (2004) provide a meta-analysis of studies of water supply, sanitation, and hygiene that updates the findings on diarrheal illness by Esrey et al. (1991). While diarrheal illness is generally not as serious as some other waterborne illnesses, it is more common and affects a larger number of people.

Water, sanitation, and hygiene factors also influence child mortality. Esrey et al. (1991) find in their review of studies that the median reduction in child mortality from improved water and sanitation was 55 percent. The term "improved water and sanitation" refers to a change from the status quo to a situation where the Millennium Development Goals (MDGs) that define improved water and sanitation are being met. Shi (1999) provides econometric estimates of the impact of potable water and sewerage connection on child mortality using a data set for about 90 cities around the world. Literacy and education level are also found to be important for parental protection of child health against environmental risk factors. Esrey and Habicht (1988) report from a study in Malaysia that maternal literacy reduces child mortality by about 50 percent in the absence of adequate sanitation, but only by 5 percent in the presence of good sanitation facilities. Literacy is also found to reduce child mortality by 40 percent if piped water is present, suggesting that literate mothers take better advantage of water availability for hygiene purposes to protect child health.

Findings from demographic and health surveys around the world further confirm the role of literacy in child-mortality reduction. Rutstein (2000) provides a multivariate regression analysis of infant and child mortality in developing countries using Demographic and Health Survey data from 56 countries from 1986 to1998. The study finds a significant relationship between infant and child mortality rates and piped water supply, flush toilet, maternal education, access to electricity, medical services, oral rehydration therapy (ORT), vaccination, dirt floor in household dwelling, fertility rates, and malnutrition. Similarly, Larsen (2003) provides a regression analysis of child mortality using national data for the year 2000 from 84 developing countries representing 95 percent of the total population in the developing world.

Estimating Incidence

The estimation of the incidence of disease in India was based significantly on the India National Family Health Survey (NFHS-3), which provides data on diarrheal prevalence in children under the age of five years. It reports a diarrheal prevalence (preceding 12 days) rate of 8.9 percent in urban areas and 9 percent in rural areas. This rate is used to estimate annual episodes per child under five, and then total annual cases in all children under five. The procedure applied is to multiply the two-week prevalence rate by 52/2.5 to arrive at an approximation of the number of annual cases of per child. The prevalence rate is not multiplied by 26 two-week periods (i.e., 52/2), but multiplied by 52/2.5 because the average duration of diarrheal illness is assumed to be three-to-four days. This implies that the two-week prevalence captures a quarter of the diarrheal prevalence in the week prior to and a quarter in the week after the two-week prevalence period.

The NFHS-3 household survey does not (nor does any other household survey in India) provide information on diarrheal illness in the population above five years of age. International experience provides an indication of the annual incidence of diarrhea per child relative to annual incidence for the rest of the population. International experience suggests that diarrheal incidence in the population above five years of age is 20 percent of incidence in children under five years. It should be noted however that usually the databases are for cases of diarrhea treated at health facilities. In general, the percentage of cases of diarrhea that are treated at health facilities is higher among young children than older children and adults. Thus, 20 percent is likely an underestimate of diarrheal cases in the population above five years of age. The annual cases of diarrhea per person among the population above five years of age, presented in Table 2.9, is therefore estimated in the range of 0.2 to 0.3 per annual case of a child under five (see Krupnick, Larsen, and Strukova, 2006).

Sometimes diarrheal illness requires hospitalization. The National Sample Survey Organization (2004) provides some information on diarrheal hospitalization in urban and rural areas. A hospitalization rate of 0.15 percent for children under five and 0.3–0.6 percent for the population over five was applied to all cases of diarrhea estimated above.

In addition to the number of cases, we also need the DALYS associated with the cases. In order to calculate these we require the disability weight for diarrheal morbidity, which is taken as 0.119 for children under five and 0.086 for the rest of the population The duration of illness is assumed to be seven days for children (as in Krupnick, Larsen, and Strukova, 2006) and from three to four days for adults.

For typhoid, the disability weight is estimated at 0.2. Duration of illness is estimated from a study in India (Sinha et al., 1999). Average duration is about 11 days for children under five and 13 days for people over five (average age is 10 years). Annual incidence of typhoid in 2009 is reported by Indiastat.com.[7]

However, the DALYs per 100,000 cases of diarrheal illness are much higher for the population over five years of age. This is because DALY calculations involve age weighting that attaches a low weight to young children and a higher weight to adults that corresponds to physical and mental development stages.[8] For diarrheal and typhoid mortality the number of DALYs lost is 34 for children under five, and 32 for those over five (typhoid mostly effects those under 14 years old). This reflects an annual discount rate of 3 percent of life years lost.

Baseline Cost Data

The baseline cost data are given in Table A1.6, with the source of the estimate in each case. Two points to note are the following:

1. Percent of diarrheal cases in the age group older than five years treated at medical facilities is estimated from percent of treated cases among children using international experience.
2. The value of time for adults is based on national average wages. Economists commonly apply a range of 50–100 percent of average urban and rural wage rates to reflect the value of time. The daily rate of Rs 150 in urban areas and Rs 60–75 in rural areas reflects around 75 percent of average weighted wage in India.[9] These rates for value of time are applied to both income-earning and nonincome-earning adults. There are two reasons for applying the rates to nonincome-earning adults. First, most nonincome-earning adults provide a household function that has a value. Second, there is an opportunity cost to the time of nonworking individuals, because they could choose to join the paid labor force.[10]

Averting Expenditures

The elements in the calculation of averting expenditures are the following.

Table A1.6 Baseline data for cost estimation: Inadequate Water, Sanitation, and Hygiene

	Baseline	*Source*
Percent of diarrheal cases treated at medical facilities (children under five years) and with medicines	58–65%	NFHS-3
Percent of diarrheal cases treated with oral rehydration solution (ORS) (children under five years)	37–44%	NFHS-3
Percent of diarrheal cases treated at medical facilities (population over five years) and with medicines	40–50%	Estimated from a combination of international experience and Krupnick et al. (2006).
Average cost of doctor visits, urban and rural (Rs)	100–500	Estimated from a combination of international experience (WHO) and per consultations with pharmacies, medical service providers, and health authorities
Average cost of medicines for treatment of diarrhea (Rs)	100	
Average cost of ORS per diarrheal case in children (Rs)	15–30	
Average duration of diarrheal illness in days (adults and children)	3–7	Krupnick et al. (2006)
Hours per day of care giving per case of diarrhea in children	2	Assumption
Hours per day lost to illness per case of diarrhea in adults	2	Assumption
Value of time for adults, care giving and ill adults (Rs per hour)	9–19	Based on urban and rural wages in India (see outdoor air pollution section)
Hospitalization rate (% of all diarrheal cases): children under five years	0.75%	NSS (2004)
Hospitalization rate (% of all diarrheal cases): population over five years	0.5%	
Average length of hospitalization (days)	2	Adjusted from Larsen (2004)
Time spent on visitation (hours per day)	4	Assumption
Average cost of hospitalization (Rs per day)	600–980	NSS (2004)
Percent of diarrheal cases attributable to inadequate water, sanitation, and hygiene	90%	WHO (2002b)

Bottled Water. From a combination of Jethoo and Poonia (2011) and NFHS-3 it was estimated that about 1.5 billion liters of bottled water are sold in urban areas and 0.5 billion liters in rural areas of India annually. These figures are used as a lower bound of bottled water consumption in India. The Worlds Water Institute provides a much higher estimate at about 40 billion liters in total.[11] We consider this as a higher bound of bottled water consumption in India. Total annual cost of bottled water consumption is estimated at about Rs 20 billion in urban areas and Rs 7 billion in rural areas.[12]

It should be noted that a portion of bottled water consumption is not only associated with perceptions of health risk of water supply, but rather also a matter of lifestyle choice and convenience. In the absence of data, no adjustment has been made to account for this. The estimated cost of bottled water consumption associated with health-risk perceptions is therefore an unknown overestimate of health-risk costs.

Boiling of Water. According to NFHS-3, 16 percent of urban households and 8 percent of rural households boil their drinking water either all the time or sometimes. Table A1.7 presents the estimated annual cost of boiling water for those households, totaling Rs 4.5–9.5 billion per year. Table A1.8 gives some of the baseline information that goes into making the calculations. It is assumed that the average daily consumption of drinking water per person is 0.5–1.0 liters among households boiling water. Residential cost of energy is estimated based on data from local experts. The average stove efficiency is for electric, natural gas, and kerosene. Lower efficiency was applied for wood stoves.

Water Filtering. According to NFHS-3, 13 percent of households filter their drinking water in urban areas and 3 percent filter drinking water in rural areas. In total about 10 million households

Table A1.7 Estimated annual cost of boiling drinking water

Type of energy	*Estimated annual cost (billion Rs)*	
	Low	High
Fuel wood	1.5	3
Kerosene	0.5	1
Natural gas	2.5	5
Other types of energy	0.0	0.5
Total annual cost	4.5	9.5

Table A1.8 Baseline data for cost estimation of boiling water

	Data	Note
Percentage of households that boil their drinking water	8–16%	NFHS-3
Average daily consumption of drinking water	0.5–1.0	Liters per person per day
Percent of households using fuel wood for cooking	32–90%	NFHS-3
Percent of households using kerosene for cooking	1–8%	
Percent of households using natural gas for cooking	9–59%	
Percent of households using other types of energy for cooking	0–1%	
Energy requirement of heating of water (100 percent efficiency)	4,200	Joules/ltr/1° C
Average stove efficiency for heating of water	50%	Varies by type of stove
Average wood stove efficiency for heating of water	20%	
Average time of boiling water (after bringing water to boiling point)	10	Minutes
Average cost of liquefied petroleum gas (LPG)	310	Rs per 14.2 kg
Average cost of kerosene	12	Rs per liter
Average cost of fuel wood	1	Rs per kg*

* Fuel wood price taken from http://infochangeindia.org/poverty/features/the-multi-billion-dollar-fuelwood-trade-is-the-last-resort-for-indias-poor.html (June 2003).

filter water in urban areas and 5 million in rural areas. With an average filter price at about Rs 4,000 and Rs 1,000 for a candle filter (per consultations with local experts), total annual filtering costs are at about Rs 14 billion in urban and Rs 7 billion in rural areas.

A1.3 Indoor Air Pollution

Desai, Mehta, and Smith (2004) provide a review of research studies from around the world that have assessed the magnitude of health effects from indoor air pollution from solid fuels and present odds ratios for acute respiratory illness (ARI) and chronic obstructive pulmonary disease (COPD). The odds ratios represent the risk of illness for those who are exposed to indoor air pollution compared to the risk for those who are not exposed. The exact odds ratio depends on several factors, such as concentration level of pollution in the indoor environment and the amount of time individuals are exposed to the pollution. A range of "low" to "high" ratios is therefore presented in Table 2.11 that reflects the review by Desai, Mehta, and Smith (2004).

Studies from around the world have also found linkages between indoor air pollution from traditional fuels and increased prevalence of tuberculosis and asthma. It is also likely that indoor air pollution from such fuels can cause an increase in ischemic heart disease and other cardiopulmonary disorders. As discussed in the section on urban air pollution, Pope et al. (2002) and others have found that the largest effect of urban fine particulate pollution on mortality is for the cardiopulmonary disease group. As indoor smoke from traditional fuels is high in fine particulates, the effect of indoor air pollution on these diseases might be substantial. More research, however, is required in order to draw a definite conclusion about the linkage and magnitude of effect.

Details of Estimation of Parameters

Annual new cases of ARI and COPD morbidity and mortality (Di) from fuel wood smoke were estimated from the following equation:

$$Di = \text{PAR} \times DiB \qquad (3)$$

where DiB is baseline cases of illness or mortality, i (estimated from the baseline data in Table A1.9), and PAR is given by:

$$\text{PAR} = PP^*(OR - 1)/(PP^*(OR - 1) + 1) \qquad (4)$$

where PP is the percentage of population exposed to fuel wood smoke (32 percent of urban and 86 percent of rural population according to India Census, 1998), and OR is the odds ratios (or relative risk ratios) presented in Table 2.11.

WHO (Desai, Mehta, and Smith, 2004) suggests to use a ventilation coefficient of 0.25 for households that use improved stoves or have kitchens outside. The national survey of solid fuel use reported in NFHS-3 estimated that kitchens are located outside of the house in 22 percent of rural households and 9 percent of urban

Table A1.9 Baseline data for cost estimation: Indoor air pollution

	Urban	*Rural*	*Source*
Percent of ARI cases treated at medical facilities (children under five years)	78.1%	66.3%	NFHS-3
Cost of medicines for treatment of acute respiratory illness	240	240	Per consultations with pharmacies
Percent of ARI cases treated at medical facilities (females over 30 years)	35%	29%	International experience
Percent of COPD patients being hospitalized per year	1.5		Assumption based on Schulman et al. (2001) and Niederman et al. (1999)
Percent of COPD patients with an emergency doctor/hospital outpatient visit per year	15		
Average number of doctor visits per COPD patient per year	1		
Estimated lost workdays (including household work days) per year per COPD patient	2.6		Estimated based on frequency of doctor visits, emergency visits, and hospitalization
Cost of doctor visit (Rs per visit)	700	100	NSS (2004), and per consultations with pharmacies and medical service providers
Cost of hospitalization (Rs per day)	980	600	
Cost of emergency visit (Rs)	800	300	
Average duration of ARI in days (children and adults)	7		Assumption based on literature
Hours per day of caregiving per case of ARI in children	2		Assumption based on literature
Hours per day lost to illness per case of ARI in adults	3		Assumption based on literature
Value of time for adults, caregiving and ill adults (Rs per hour)	19	7.5–9.5	75% of rural wages in India
Average days hospitalization for COPD	10		Larsen (2004b)

households. The solid fuel use exposure formula was corrected accordingly.

The following details relate to Table 2.13.

1. WHO estimates of COPD mortality for India are utilized in the analysis. COPD morbidity incidence, according to international disease classifications, are not readily available for India. Regional estimates from WHO and Shibuya, Mathers, and Lopez (2001) for the Sear D-WHO subregions are therefore applied.
2. As in NFHS-3, the national average two-week prevalence rate of ARI in children under five years is used to estimate total annual cases of ARI in children under five. The procedure applied is to multiply the two-week prevalence rate (24 percent) by 52/3 to arrive at an approximation of the annual cases of ARI per child. A factor of 52/3 is applied because the average duration of ARI is assumed to be about seven days. This implies that the two-week prevalence captures half of the ARI prevalence in the week prior to and the week after the two-week prevalence period.
3. There is no information on ARI prevalence in adults. Krupnick, Larsen, and Strukova (2006) provide an indication of the annual incidence of ARI per child relative to annual incidence for the rest of the population. An analysis of the database suggests that ARI incidence in the population above five years of age is 0.36 of the incidence in children under five years. In general, the percentage of cases of ARI that are treated at health facilities is higher among young children than older children and adults. For instance, in Krupnick, Larsen, and Strukova (2006), the percentage of treated cases of ARI among children under four years old is 1.15 times higher than among four-year-old children. Thus the incidence ratio of 0.36 is likely an underestimate of ARI cases in the population above five years of age. The annual cases of ARI per person among the population above five years of age, presented in Table 2.13, is therefore estimated in the range of 0.36 to 0.42 [(1/(0.85))*0.36] of the annual cases per child under five.
4. ARI mortality in children under five years is presented in Table 2.13. Twenty-two percent of the mortality of children under age 5 is due to respiratory infections as reported by

the Office of Registrar General and Commissioner of Census (2004). It suggests a high mortality load among the corresponding category of population in India.

5. Table 2.12 also presents DALYs per cases of ARI and COPD, which are used to estimate the number of DALYs lost because of indoor air pollution. The disability weight for ARI morbidity is the same for children and adults (i.e., 0.28), and the duration of illness is assumed to be the same (i.e., seven days). The DALYs per 100,000 cases of ARI is however much higher for adults. This is because DALYs calculations involve age weighting that attaches a low weight to young children and a higher weight to adults, corresponding to physical and mental development stages.[13] For ARI child mortality the number of DALYs lost is 34. This reflects an annual discount rate of 3 percent of life years lost.
6. DALYs lost per case of COPD morbidity and mortality is based on life tables and age-specific incidence of onset of COPD reported by Shibuya, Mathers, and Lopez (2001) for the Sear D region. A disability weight of 0.2 has been applied to COPD morbidity. A discount rate of 3 percent is applied to both COPD morbidity and mortality.

Baseline Data for Costs

Baseline data for the cost estimates of morbidity are given in Table A1.9. The percentage of ARI cases in the age group older than five years treated at medical facilities is estimated from percent of treated cases among children (NFHS-3) and the ratio of treated cases among children under five to treated cases among the population over five years of age (Krupnick, Larsen, and Strukova, 2006).

The value of time for adults is 75 percent of urban and rural average daily wages, resulting in rates of Rs 150 and Rs 60–75, respectively. The rationale for valuation of time was discussed in the sections on water, sanitation, and hygiene and on urban air pollution.

There is very little information about the frequency of doctor visits, emergency visits, and hospitalizations for COPD patients in any country in the world. Schulman and Bucuvalas, Inc. (2001) and Niederman et al. (1999) provide some information on this from the United States and Europe. Figures derived from these studies are applied to India in this report. Estimated lost workdays per year is based on frequency of estimated medical treatment plus an additional seven days for each hospitalization and one extra day for each doctor and emergency visit. These days are added to reflect time needed for recovery from illness.

To estimate the cost of a new case of COPD, the medical cost and value of time losses have been discounted over an assumed 20-year duration of illness. An annual real increase of 2 percent in medical cost and value of time has been applied to reflect an average expected increase in annual labor productivity and real wages. The costs are discounted at 3 percent per year, a rate commonly applied by WHO for health effects.

A1.4 Valuation of Premature Mortality

Two distinct methods of valuation of premature mortality are commonly used to estimate the social cost of premature death: the human capital approach (HCA) and the value of statistical life (VSL). The first method involves estimating income losses from premature death and has been dominant in the past. Because this measure is not based on individual preferences and for other conceptual problems, it has been overtaken by both stated and revealed preference approaches to estimating preferences for reducing mortality risks. The monetary value of these preferences, or willingness to pay, when divided by the relevant risk reduction yields the value of statistical life (VSL). Because HCA almost always underestimates the VSL, the HCA has been applied as a low estimate and VSL as a high estimate in estimating the cost of premature mortality.

Human Capital Approach

The HCA is based on the economic contribution of an individual to society over the lifetime of the individual. Death involves an economic loss that is approximated by the loss of all future income of the individual. Future income is discounted to reflect its value at the time of death. The discount rate commonly applied is the rate of time preference. Thus the social cost of mortality, according to the HCA, is the discounted future income of an individual at the time of death. If the risk of death, or mortality risk, is evenly distributed across income groups, average expected future income is applied to calculate the social cost of death. Mathematically, the present value of future income is expressed as follows:

$$PV_0(I) = \sum_{i=k}^{i=n} I_0(1+g)^i / (1+r)^i \quad (5)$$

where PV_0 (I) is present value of income (I) in year 0 (year of death), g is annual growth in real income, and r is the Ramsey discount rate. As can be seen from equation (5), the equation allows for income to start from year k and end in year n. In the case of children, we may have i {16, . . . ,65}, assuming the lifetime income on average starts at age 16 and ends at retirement at age 65. The annual growth of real income is variable and set at about 5 percent for the first 30 years, reducing to 2 percent over the next 35 years. The GDP per capita growth rate was computed in the CGE model for India (see the World Bank's forthcoming report, "Economic Growth and Environmental Sustainability"). Since the real growth of GDP per capita is quite high, it should be accounted for in determining the social discount rate. We apply the Ramsey discount rate to real GDP per capita, assuming an intertemporal coefficient of consumption equal to 1, as in Summers and Zeckhauser (2008). Then the average effective discount rate is set at about 1 percent.

The most important practical issue raised regarding the HCA is the application of this valuation approach to individuals that do not participate in the economy—that is, to individuals who do not have an income, such as the elderly, family members taking care of the home, and children. One may think of an extension of the HCA that recognizes the value of nonpaid household work at the same rate as the average income earner or at a rate equal to the cost of hiring a household helper. In this case, the HCA can be applied to nonincome earners and children (whether or not children will become income earners or take care of the home during their adult life). In the case of the elderly, the HCA would not assign an economic value to individuals who have either retired from the workforce or do not make significant contributions to household work. This obviously is a serious shortcoming of the HCA approach.

Table A1.10 Cost of mortality per death using the Human Capital Approach (HCA)

Cause of mortality	*Average number of years lost*	*Rs (thousand)*
Adults		
Urban air pollution	8	430
Indoor air pollution	7	390
Lead exposure	8	430
Children		
Urban air pollution	65	1,148
Indoor air pollution	65	1,148
Diarrheal illness and typhoid, children under 5	65	1,148
Diarrheal illness and typhoid, children under 19	55	1,863

The estimated cost of mortality in India based on HCA is presented in Table A1.10. Average annual income is approximated by GDP per capita, corresponding to around Rs 57,000 per year. The estimates are from equation (5).

Value of Statistical Life

While the HCA involves valuation of the death of an individual, VSL is based on preferences for reducing mortality risk by a small amount. Everyone in society is constantly facing a certain risk of dying. Examples of such risks are occupational fatality risks, risks of traffic accident fatality, and environmental mortality risks. It has been observed that individuals adjust their behavior and decisions in relation to such risks. For instance, individuals may demand a higher wage (a wage premium) for a job that involves a higher than average occupational risk of fatal accident, purchase safety equipment to reduce the risk of death, or be willing to pay a premium or higher rent for properties (land and buildings) in a cleaner and less polluted neighborhood or city.

Through the observation of individuals' choices and willingness to pay for reducing mortality risk, it is possible to measure or estimate the value to society of reducing mortality risk or, equivalently, measure the social cost of a particular mortality risk. For instance, it may be observed that a certain health hazard has a mortality risk of 1/10,000. This means that one individual dies every year (on average) for every 10,000 individuals. If each individual on average is willing to pay Rs 10 per year for eliminating this mortality risk, then every 10,000 individuals are collectively willing to pay Rs 100,000 per year. This amount is the VSL. Mathematically it can be expressed as follows:

$$VSL = WTPAve \times 1/R \qquad (6)$$

where WTPAve is the average willingness-to-pay (rupees per year) per individual for a mortality risk reduction of magnitude R. In the previous illustration, $R = 1/10,000$ (or $R = 0.0001$) and WTPAve = 10 rupees. Thus, if 10 individuals die each year from the health risk illustrated earlier, the cost to society is 10* VSL = 10* Rs 100,000 = Rs 1 million.

A number of VSL studies have been conducted in India. Table A1.11 presents a summary.

Table A1.11 Value of statistical life in India

312 pt	*Method of estimation*	*Value (Rs)*	*Adjusted value (Rs, 2010)*	*Adjusted value (US$, 2010)**
Shanmugam (1997)	Compensating-wage differentials	12,084,000	18,932,020	$398,569
Simon et al. (1999)	Compensating-wage differentials	6,417,341–15,040,642	16,197,563	$341,001
Bussolo and O'Connor (2001)	PPP and income elasticity, using Brandon and Homman (1995) estimate	$202,000–343,860, use the central value of $273,000	19,109,280	$402,301
Madheswaran (2007)	Compensating-wage differentials	15,000,000	16,939,353	$356,618

Using an exhange rate of Rs 47.5 = US$1.

Source: Prepared by A. Sagar.

The average VSL from these comes out at about Rs 17.8 million (US$375,000), and this figure was applied in the report. In this report we used the average of the VSL and HCA values for adults (i.e., US$192,000 or Rs 9.1 million). For children we do not use the VSL value as none of the VSL studies are for children. Hence we take only the HCA value of US $24,168 or Rs 1.148 million (Table 6.1). This conservative approach is also consistent with other costs of degradation studies that have been conducted.

Notes

1. Data are from http://www.indiastat.com.
2. WHO member states are grouped into six geographical regions: AFRO (Africa), AMRO (Americas), EMRO (Eastern Mediterranean), EURO (Europe), SEARO (South-East Asia), and WPRO (Western Pacific). The six regions are further divided based on patterns of child and adult mortality in groups ranging from A (lowest) to E (highest): AFR (D,E); AMR (A,B,D); EMR (B,D); EUR (A,B,C); SEAR (B,D); WPR (A,B). See World Health Organization at www.who.int.
3. See National Institute of Health, Fogarty International Center, Advancing Science for Global Health, at http://www.fic.nih.gov/dcpp/weights.xl.
4. The disability weight for mortality is 1.0.
5. Data are from http://www.indiastat.com.
6. Some may argue that the value of time based on wage rates should be adjusted by the unemployment rate to reflect the probability of obtaining paid work.
7. Data are from http://www.indiastat.com.
8. It should be noted that some researchers elect not to use age weighting, or they report DALYs with and without age weighting.
9. This corresponds to a daily urban average wage rate of about Rs 200. and rural wage rate of Rs 80–100, estimated from Indiastat.com.
10. Some may argue that the value of time based on wage rates should be adjusted by the unemployment rate to reflect the probability of obtaining paid work.
11. See "Per Capita Bottled Water Consumption, by Country, 1999 to 2004" at http://www.worldwater.org/data20062007/Table13.pdf.
12. For information about price and cost of bottled water production in India, see http://www.gits4u.com/water/water16.htm supplies.
13. It should be noted that some researchers elect not to use age weighting, or they report DALYs with and without age weighting.

Appendix 2

Methodology for Estimating the Cost of Natural Resource Degradation

A2.1 Soil Degradation

There is a lot of evidence that India has a substantial land degradation problem. Official data on land degradation are summarized in Table 2.15. Total degraded area is 188 million hectares, which amounts to about 60 percent of total reporting land for land utilization statistics in the country.[1]

Salinity Losses

The estimated losses from saline soils were calculated under the assumption that such land is only used for wheat production (if it is used at all). This reflects the assumption that when soils are saline, farmers will tend to plant crops that are more tolerant of this factor, such as wheat, as opposed to pulses and rice. The FAO estimates indicate a loss of yield of 5 percent for wheat per unit salinity (dS/m) for levels of salinity over 6 dS/m. Taking these values and applying them to lands under wheat is the basis of the estimated loss of output. Research by the Central Soil Salinity Research Institute (CSSRI, 2010) estimates about three million hectares of agricultural land as saline.

Two scenarios are considered, both of which assume that the total land cultivated for wheat in saline conditions is 2.9 million hectares. In scenario 1, it is assumed that these lands are only slightly saline (EC = 4–8 dS/m). In scenario 2, some of this land is assumed to be slightly saline (2 million hectares), but some wheat is also cultivated on moderately saline lands (0.9 million hectares). The estimated losses are then multiplied by the wheat farmgate prices in 2009 (Rs 12,000 per metric ton; Food and Agriculture Organization, 2011) and the costs of production are deducted to arrive at a net loss figure.[2]

Waterlogging Losses

We assume that rice is mostly cultivated on waterlogged lands. Average rice yield losses on waterlogged land are assumed to be 40 percent of the observed yield (as in Gundimeda et al. (2005)). Based on data from Indiastat.com, it is estimated that rice is cultivated on 1.7 million hectares of waterlogged lands. Furthermore, the annual farmgate price of paddy is Rs 18,000 (Food and Agriculture Organization, 2011).

Soil Erosion Losses

The State of Environment, India (Ministry of Environment and Forests and World Bank, 2001) and Gundimeda et al. (2005) report that annually about 29–55 metric tons of major nutrients are leached out from the land in India. Table A2.1 presents an estimate of the amount of fertilizers required to substitute annual humus loss of nutrients through leaching.

A2.2 Pasture Degradation

Data on the extent of degraded grazing lands were not readily available. In the past 60 years lands available for grazing have remained relatively stable, but livestock measured in adult cattle units (ACU) have increased by about 50 percent. The impact of this increase in pressure has been a decline in the fodder available on rangelands. Based on interviews with rangeland experts in India and data in

Table A2.1 Fertilizers for nutrient loss substitution

	Amount required to replace the leached-out major nutrients, in tonnes		*Price, 2009 (Rs per tonne)*
	Gundimeda (2005)	*State of environment (2001)*	
Nitrogen	1.4	0.8	22,000
Phosphorus	3.3	1.8	15,000
Potassium	50.2	26.3	10,375

Note: Nitrogen recalculated from price of urea (46% N). Phosphorus is recalculated from price of diammonium phosphate (18-46-0). Potassium is recalculated from price of muriate of potash (60% K2O), as presented on Indiastat.com.

Source: Ministry of Environment and Forests and World Bank (2001) and Gundimeda et al. (2005); indiastat.com.

Roy and Roy (1996),[3] the current average yield is estimated at 1.1 metric tons of dry matter (DM) per hectare on degraded rangelands. In the absence of degraded grazing land productivity data, we assume that productivity on the degraded lands is at 0.55 tons of dry matter per hectare. Original productivity is assumed at the 3.5 tons of dry matter per hectare. This is at the lower level of different grazing lands productivities presented in Roy and Roy (1996).

For the first method we use a fodder price of Rs 4,000–8,000 per ton of dry matter.[4] Based on that price, the loss from the reduced fodder production amounts to an average of Rs 400–800 billion per year, for a sustainable rangeland fodder utilization rate of 40–60 percent (see Hocking and Mattick, 1993). The loss accounted for about 0.91 percent of GDP in 2010.

Additional losses could be attributed to complete loss of pasture lands and their transfer to barren lands. However, there are no reliable data that would allow estimation of this loss.

The second method takes the loss in fodder and calculates the number of animals it would support and the net income from these animals. The steps in the calculations are as follows:

1. Due to degradation the fodder from the rangelands in India has declined by between 89 and 134 million tons (TDM). This is based on a rangeland area of 79.8 million hectares, with a sustainable consumable yield of between 1.4 and 2.1 tons per hectare. Due to degradation this yield has fallen by 80 percent.
2. The decline in yield could have supported 50 and 75 million ACUs, given that each ACU requires 1.8 tons of dry matter per annum.
3. Each ACU contributes Rs 3,410 to the GDP. This is based on the fact that there are 499 million ACUs in India and their total contribution to GDP is Rs 1,702 billion.
4. Hence the total loss in income from the degradation is between Rs 170 billion and Rs 256 billion, or between 0.3 and 0.4 percent of GDP.

A2.3 Forest Degradation

Loss of forest value by the degraded forest is at the core of forest degradation methodology. The methodology for forests valuation is presented below. Only forest use values are estimated in the report.

The use values included in the study are taken from the extended study by the Green Accounting for Indian States and Union Territories Project (GAISP) (2005–2006), which was designed to build a system of adjusted national accounts for India to estimate genuine national wealth as a comprehensive measure of growth instead of GDP. We applied some of the estimates developed in this study to estimate forest degradation in India.

Forests yield a wide variety of plants and animals used in the traditional lifestyle and farming system: (a) foodstuffs, including mushrooms, fruits, nuts, roots, game, and leaves, to complement diets or generate small amounts of cash; (b) medicinal plants and seasonings, either used domestically or sold in local markets; (c) construction materials and materials for household utensils, including furniture wood, roofing materials, mats, trays, storage containers, and house timber; (d) fuel wood for cooking and small-scale enterprises; and (e) commercial extraction of chicle and resins. Forests are also supplementary areas for grazing and, in the tropical zones, are used in rotation in the traditional slash-and-burn agricultural systems. Direct use values for forest lands could be estimated based on direct market values of goods produced there. Values of major forest goods, like roundwood (including timber and fuel wood), nontimber values, and fodder were estimated using market prices. The World Bank (2006) reports that 5–42 percent of rural household income is generated by forest products.

Logging

Food and Agriculture Organization's Forest products (FAO, 2009) reported annual roundwood production in India. It includes all wood removed with or without bark, including wood removed in its round form, split, roughly squared, or in other form (e.g., branches, roots, stumps, and burls, where these are harvested). Fuelwood is included in this aggregate. FAO (2009) estimates that annual roundwood production is at about 3.3 million cubic meters. If as in Gundimeda (2005) about 10 percent of forest is destroyed at the time of logging, then total roundwood removed is at about 3.7 million cubic meters annually. Brandon and Homman (1995) suggests an average stumpage price of US$100 per cubic meter of roundwood. This price approximately corresponds to the roundwood profit margin reported by the World Bank (2006). Following from that, the estimated value of annual timber extraction in India is about Rs 17 billion.

Nontimber Value

As in Gundimeda (2005), we apply a conservative estimate for nontimber values at Rs 301 per hectare. Nontimber value is therefore estimated at about Rs 21 billion annually.

Fodder

Fodder is estimated in Gundimeda (2005) at about 23.6 million tons annually (4.9 tons of dry matter, and 0.1 tons of leaf biomass per hectare). With current fodder prices at Rs 8,000 per ton and relatively cheap substitution of straw at Rs 4,000 per ton, total value of fodder generated in forest cover land is in the range of Rs 94–189 billion.

Recreations Use (Ecotourism)

Gundimeda (2005) applies a travel-cost method to estimate ecotourism value per hectare of Indian forests. Only national parks are assumed to attract tourists. She estimates that 15.7 million hectares of natural parks in India bring Rs 14,165 million annually. Assuming growth of the tourist industry up to 2020, annualized net present value of the tourist industry (at a 4 percent discount) Rs 3,260 per hectare, or about Rs 51 billion total for all natural parks. This estimate reflects potential benefits from forests in India, so this benefit is quite uncertain.

Carbon Sequestration

Carbon storage is another important function of forests that adds to its value. Many studies estimate the carbon potential of Indian forests. We apply estimated average net carbon accumulation by hectare of forest as reported in Gundimeda (2001) at 1.1–1.4 tons per hectare annually. The carbon price is assumed at US$20 per ton of CO_2, which corresponds to the recent estimate of the social price of carbon. Other alternatives for carbon price (e.g., CDM price) do not provide a viable alternative. For instance, CDM price is mainly driven by European Union's Emission Trading Scheme (EU ETS) limits on international offsets. The EU-regulated entities have nearly filled their limits (including phase 3 EU ETS), and therefore, over the last several months the spread between

EU Allowances and Certified Emission Reduction under the Clean Development Mechanism continues to grow.

Annual benefits from net carbon accumulation by forests are in the range of Rs 270–340 billion.

Indirect Use Values

Indirect use values of forests include watershed protection, nutrient-loss prevention, erosion and flood prevention, and water and nutrient recycling. Although there is no definite agreement in the literature about the magnitude of this forest value, Pearce, Putz, and Vanclay (1999) present a higher end estimation of US$30 per hectare of forest, generalized from the literature review. In this study, erosion prevention value was estimated using data on nutrient-loss prevention per hectare of dense forest as reported in Gundimeda (2005). Total soil loss prevented by dense forest is estimated using fertilizer prices from Indiastat.com and India Development Gateway,[5] recalculated per ton of N, P, and K. The resulting figure of total soil loss prevented by forests comes out at Rs 15.5 billion. Details are given in Table A2.2.

Water recharge value was estimated from the opportunity price of water adjusted to 2009 with the wholesale price index (Rs 7.9 per cubic meter). Water recharge value per hectare of dense forest was estimated in Gundimeda (2005) at Rs 106 per cubic meter. Total water recharge value of the forest is Rs 6.4 billion. Flood prevention damage was excluded from consideration since flood losses were estimated in a separate chapter. Gundimeda (2005) suggested that presence of dense forest will reduce flood damage by one third. However, separate analysis is required to associate flood damage function with dense forest in each state where dense forest is present.

Table A2.2 Estimation of erosion loss prevention function in India

	N (urea)	*P (diammonium phosphate)*	*K (muriate of potash)*	*Organic carbon*
Loss prevented (tonnes)	232,492	4,409	826,749	2,254,770
Effective price (Rs per tonne)	25,000	6,880	10,375	500
Loss prevented (billions Rs)	5.8	0.03	8.6	1.1

A2.4 Natural Disaster Costs

India is annually afflicted by such natural disasters as floods, landslides, tropical cyclones, and storms. The total mean annual cost of natural disasters was estimated at Rs 150 billion, or 0.23 percent of GDP in 2009.

In the literature, floods and storms, including tropical cyclones, are indicated as a significant source of natural hazard and damage for human health, agriculture, real estate, infrastructure, and personal property. The occurrence of natural disasters is highly uncertain. Although published data are incomplete and very often not comparable, based on available sources it is possible to analyze implied damage from natural disasters in India. Figure A2.1 presents occurrence of natural disasters, including floods, heat and cold waves, storms, tropical cyclones, droughts, mudslides and landslides, epidemics, and so on registered by EM-DAT starting in 1980 in India.

During the last decade the frequency of natural disasters has significantly increased. Climate change could account for some of this increase, but anthropogenic activity was an important confounding factor that exacerbated the negative impact of natural causes. The main types of natural disasters were tropical cyclones, floods, and severe storms, and their frequency increased over the last decade (see Figure A2.2).

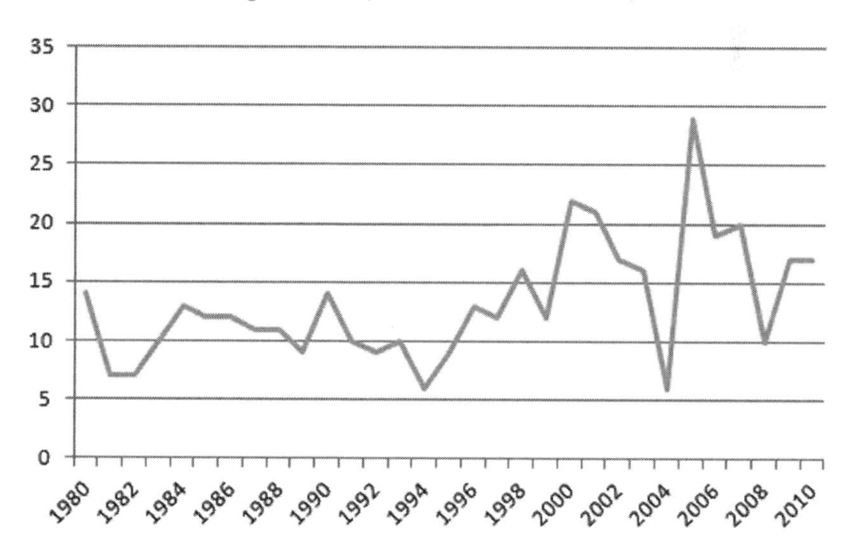

Figure A2.1 Annual occurrence of natural disasters in India, last 30 years

Source: EM-DAT-International Disaster Database (2011).

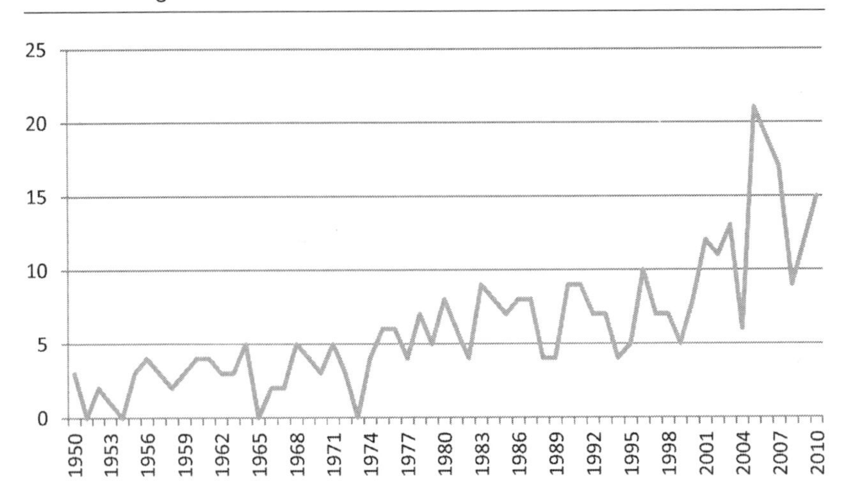

Figure A2.2 Floods, tropical cyclones, and storms: Annual occurrence in India in the past 60 years

Source: EM-DAT-International Disaster Database (2011).

Impacts of Natural Disasters

Economic losses from natural disasters include direct and indirect costs. Direct costs include human health losses in terms of mortality and morbidity, property damage, crop and livestock damage in agriculture, and public infrastructure losses. Due to the lack of data we were not able to estimate indirect losses, reflected in the contraction of economic activity, property value losses, and so forth that are associated with both short-term and long-term shocks to the economy induced by natural disasters. Figure A2.3 presents estimated direct losses from floods and storms in India starting 1993.

The direct economic losses from natural disasters were estimated using physical indicators of losses due to floods and heavy rains as presented in Indiastat.com. Details are given in Table A2.3.

Figure A2.3 presents the composition of annual damage associated with natural disasters, estimated at an average of Rs 150 billion per year over the period 1953–2009 (in constant 2009 prices).[6] As a percentage of GDP, we look at damages over the relatively recent past, as the level of damages is a function of the level of development. At the same time figures for one year can be misleading as disasters have a high degree of volatility. Hence we consider

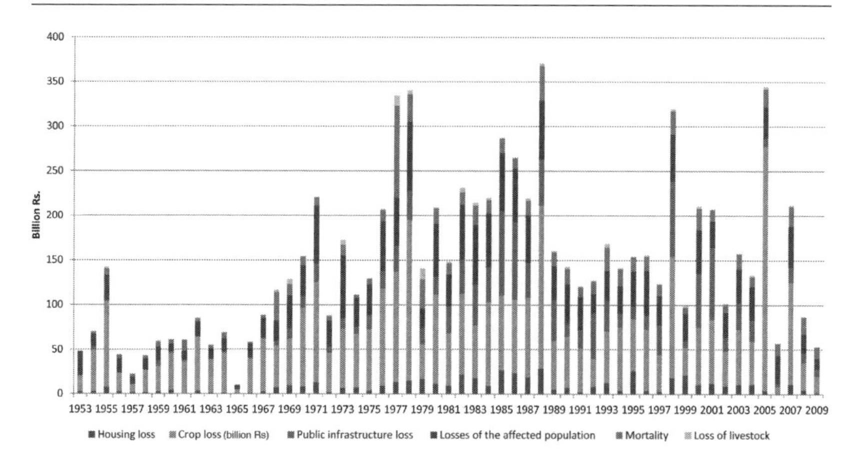

Figure A2.3 Composition of annual damage associated with natural disasters

Source: Staff estimates.

Table A2.3 Methods of valuation for natural disasters

Category of damage	*Method of valuation*	*Comments*
Loss of life	Average of HCA and VSL approaches for adults; HC approach for children	Rs 9 million per adult Rs 1.4 million. per child
Injury	Based on loss of earnings for 0.5 month per event at 75 percent of wages	Rs 1,100 per person per event
Crops	Loss of net revenue per hectare under wheat and paddy with cropping intensity of 1.39	Average net revenue for wheat and paddy was taken as Rs 13,000 per hectare
Livestock	Market price of an indigenous cow	Priced at Rs 20,000, per expert estimate
Property damage	Adjusted for inflation information from Indiastat.com	Adjustment based on WPI from Indiastat.com
Public infrastructure	Adjusted for inflation information from Indiastat.com	Adjustment based on WPI from Indiastat.com

Source: Staff estimates.

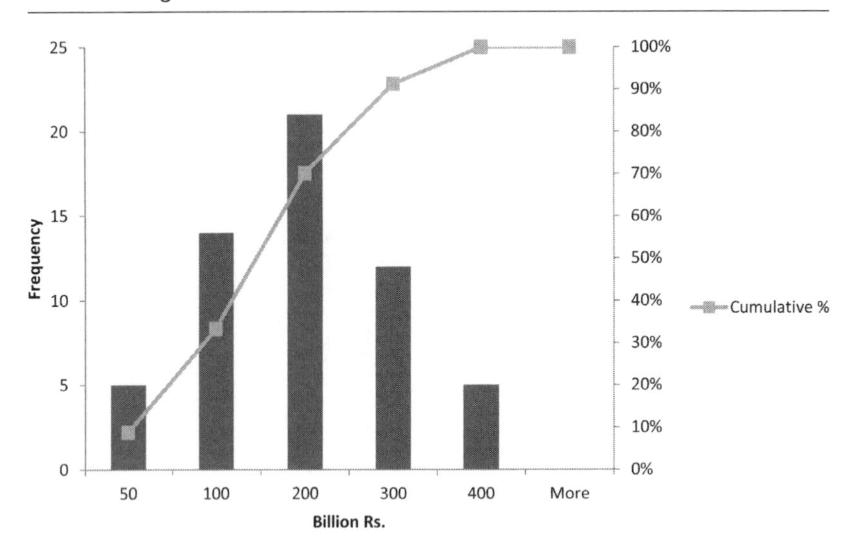

Figure A2.4 Histogram of annual losses distribution

Source: Staff estimates.

the average damages from 2000 to 2009, which turn out to be 0.37 percent of 2009 GDP. Crop losses dominate the total damage (about 45 percent of average losses) followed by losses to the affected population (about 24 percent of damage).

Estimated annual damage exhibits substantial variability and slight tail risk of occurrence of rare events with high negative outcomes. Figure A2.4 presents a histogram of annual loss distribution over the period 1953–2009. The most likely magnitude of annual damage is in the range of Rs 100–300 billion, with a mean at Rs 150 billion and standard deviation (SD) at Rs 87 billion. Damage distribution has a slightly positive skew and kurtosis. It confirms a conclusion about right tail: low frequency of events with high anticipated losses from natural disasters. In the future, frequency and value of these events may increase.

The natural disaster analysis aims to demonstrate the magnitude of economic losses related to natural disasters. These losses are not entirely attributable to environmental degradation, but attribution of this damage to different anthropogenic and nonanthropogenic causes was not in the scope of this study. Yet the costs related to natural disasters are seen as environment-related and are generally higher where protection measures are limited—typically in developing countries. Moreover, there has been an increase in damages arising from such disasters over the past decades. Hence information on trends on damages from natural disasters could be of interest.

The estimate was based on information available and performed in a conservative way. Results of the analysis for each type of disaster are aggregated and averaged. Although a distinction between flow and stock is important, for housing and infrastructure losses we included the full recovery cost. If houses are destroyed regularly, then the recovery cost appears as a flow. For more detailed analysis, the methodology developed under the auspice of the Global Facility for Disaster Reduction Recovery could be applied.

Notes

1. 305.67 million hectares in 2008. Data are from Indiastat.com.
2. The costs of wheat production are taken from Indiastat.com.
3. Tons of dry matter per hectare estimated from adult cattle unit (ACU) consumption of 2 percent of body weight per day.
4. Price of grains residuals and grass fodder from http://www.downtoearth.org.in/node/802.
5. See www.indg.in.
6. Losses starting from 1953 are valued at 2009 prices since some components of the losses were not valued (loss of life, losses of affected people, livestock loss). Crop losses were estimated in the same way to maintain comparability of the cost components. Other losses from Indiastat.com (housing losses and public infrastructure losses) were adjusted for inflation since there is no data on the level of assets loss.

Appendix 3

Details of the Meta-Analysis Functions Used in Estimating Value of Ecosystem Services

A3.1 Passive Use of Forests

Based on an extensive review of the literature, Chiabai et al. (2011) identified 27 usable estimates of forest passive use values.[1] The studies were conducted mainly in Europe and North America, although there were also studies from Brazil, China, Madagascar, and Israel. The estimated benefit transfer function is:

$$V_{WR} = V_{Eu}^{*}\left(\frac{N_{WR}}{N_{Eu}}\right)^{\delta}\left(\frac{S_{Eu}}{S_{WR}}\right)^{\sigma}\left(\frac{PPPGDP_{WR}}{PPPGDP_{Eu}}\right)^{\gamma}$$

The notations *WR* and *Eu* denote figures referring to, respectively, the WR_{-th} world region and the study site Europe region. V_{WR} is the estimated willingness-to-pay (WTP) stock value per hectare in the WR_{-th} world region. *S* denotes the forest area designated to conservation in the relevant region. *N* denotes the population of each region, and *PPPGDP* indicates the GDP adjusted using PPP taken from the *World Development Indicators*.

The estimated values for the coefficients δ, σ, and γ were 0.64, –0.39, and 0.74 respectively. All three were statistically significant.

A3.2 Grasslands[2]

The mean of grasslands value in the original studies was US$216 per hectare per year, and the median was US$37 per hectare per year. These values are low in comparison to those of the other biomes examined in the original study.

Given the very limited sample size of grassland ecosystem service values, the number of explanatory variables that can be included in the value function is also low. The explanatory variables included in the value function are GDP per capita; the area of grassland within a 50 kilometer radius of the study site; the length roads within 50 kilometers of the study site; and the accessibility index.

The estimated equation is given in Table A3.1. The estimated coefficients on the explanatory variables all have the expected signs but are only marginally statistically significant. The positive coefficient on the income variable (GDP per capita) indicates that grassland ecosystem services have higher values in countries with higher incomes; that is, grassland ecosystem services are a normal good for which demand increases with income. The negative effect of grassland abundance (area of grassland within 50 kilometer radius) on value indicates that the availability of substitute grassland areas affects the value of ecosystem services from a specific patch of grassland. The negative effect of roads on grassland values captures the effect of fragmentation on the provision of ecosystem services from grassland. Grasslands that are more fragmented by roads tend to have lower values. The positive coefficient on the accessibility index indicates that grassland areas that are more accessible tend to have higher values. In this case, direct use values derived from grasslands

Table A3.1 Grasslands valuation function

Variable name	*Variable definition*	*β*	*SE*	*Sig.*
Constant		–2.366	5.094	0.444
GDPPC_IN	Natural log of country level GDP per capita (PPP US$ 2007)	0.856	0.514	0.120
GRA50_LN	Natural log of area of grassland within 50 kilometer radius of study site	–0.029	0.142	0.839
RDS50_LN	Natural log of length of roads within 50 kilometer radius of study site	–0.225	0.213	0.309
SITES_AI	Accessibility index	2.590	1.322	0.072
N		17		
Adjusted R^2		0.27		

Source: Hussain et al. (2011).

(e.g., recreation and food provisioning) appear to dominate values that do not require access (e.g., wildlife conservation).

A3.3 Wetlands

The average wetland value is US$4,774 per hectare per year, and the median is US$250 per hectare per year. The explanatory variables included in the value function are as follows: the area of the wetland study site; the GDP per capita of the country in which the study site is located; the area of lakes and rivers within a 50 kilometer radius of the study site; the area of wetlands within 50 kilometers of the study site; the population within 50 kilometers; and the human appropriation of net primary product (HANPP) within 50 kilometers.

The value function is presented in Table A3.2. The estimated coefficients on the explanatory variables all have the expected signs and are all statistically significant at the 5-percent level, except for

Table A3.2 Wetlands valuation function

Variable name	*Variable definition*	*β*	*SE*	*Sig.*
Constant		1.708	1.978	0.725
AREA_LN	Natural log of the study site area (hectares)	–0.209	0.049	0.000
GDPPC_IN	Natural log of country level GDP per capita (PPP US$ 2007)	0.610	0.106	0.000
LAK50_LN	Natural log of area of lakes and rivers within 50 kilometer radius of study site	0.159	0.081	0.050
WET50_LN	Natural log of area of wetlands within 50 kilometer radius of study site	–0.175	0.048	0.000
POP50_LN	Natural log of population within 50 kilometer radius of study site	0.426	0.106	0.000
HAN50_LN	Natural log of human appropriation of net primary product within 50 kilometer radius of study site	–0.201	0.118	0.091
N		247		
Adjusted R^2		0.32		

Source: Hussain et al. (2011).

HANPP, which is significant at the 10-percent level. The negative effect of the area of the wetlands indicates diminishing returns to scale for wetland values. In other words, the value of an additional hectare to a large wetland is of lower value than an additional hectare to a small wetland. The positive effect of the income variable (GDP per capita) indicates that wetland ecosystem services have higher values in countries with higher incomes; that is, wetland ecosystem services are normal goods for which demand increases with income.

The positive effect of the area of lakes and rivers in the vicinity of a wetland indicates that lakes and rivers are complements to wetland ecosystem services; that is, the combination of surface water bodies results in higher value ecosystem services. The negative effect of the size of other wetland areas in the vicinity of a wetland indicates substitution effects between wetlands. The ecosystem services from a specific wetland will be of higher value if there are fewer other wetlands in the vicinity.

The positive effect of population on the value of wetland ecosystem services relates to market size or demand for those services. A larger population in the vicinity of a wetland means that more people benefit from the services it provides. The negative effect of HANPP indicates the effect of ecosystem degradation on the value of services provided by wetlands. More intensive use and appropriation of environmental resources has a negative effect on the value of wetland services.

A3.4 Mangroves

The average value of mangroves in the original studies was US$803 per hectare per year, and the median was US$220 per hectare per year. The explanatory variables included in the value function are as follows: the area of the mangrove study site; the GDP per capita of the country in which the study site is located; the population within a 50 kilometer radius of the site; the length of roads within 50 kilometers; the area of mangroves within 50 kilometers; the area of urban land use within 50 kilometers; and the area of wetlands within 50 kilometers of the study site.

The value function is presented in Table A3.3. The estimated coefficients on the explanatory variables mostly have the expected signs and are all statistically significant at the 5-percent level, except for the length of roads variable, which is significant at the

10-percent level. The negative coefficient on the area of the mangrove site indicates diminishing returns to scale. The positive effect of the income variable (GDP per capita) indicates that mangrove ecosystem services have higher values in countries with higher incomes. The positive effect of population on the value of mangrove services relates to market size or demand. A larger population in the vicinity of a mangrove means that more people benefit from the ecosystem services it provides.

The positive effect of the area of other mangroves on the value of a mangrove study site indicates that mangrove patches within a region are complementary. This suggests that isolated patches of mangrove are of lower value than more intact contiguous mangrove systems. The negative effect of the area of urban land uses in the vicinity of a mangrove reflects the associated effect of degradation

Table A3.3 Mangroves valuation function

Variable name	*Variable definition*	β	*SE*	*Sig.*
Constant		–8.239	3.157	0.010
AREA_LN	Natural log of the study site area	–0.311	0.069	0.000
GDPPC_IN	Natural log of country level GDP per capita (PPP US$ 2007)	1.499	0.218	0.000
POP50_LN	Natural log of population within 50 kilometer radius of study site	0.572	0.194	0.004
MAN50_LN	Natural log of area of wetlands within 50 kilometer radius of study site	0.208	0.083	0.014
URB50_LN	Natural log of human appropriation of net primary product within 50 kilometer radius of study site	–0.382	0.173	0.029
RDS50_LN	Natural log og length of roads within 50 kilometer radius of study site	–0.317	0.182	0.084
WET50_LN	Natural log of area of inland wetland within 50 kilometer radius of study site	–0.158	0.064	0.016
N		111		
Adjusted R^2		0.41		

Source: Hussain et al. (2011).

on the value of ecosystem services. Similarly, the negative effect of roads on mangrove ecosystem services reflects the detrimental effects of fragmentation. The negative coefficient on wetland area in the vicinity of a mangrove indicates substitution effects between wetlands and mangroves. The estimated value function is a relatively good fit with the data, with an adjusted R^2 of 0.41 showing that 41 percent of variation in mangrove values is explained by the model. This still means that 59 percent of variation in values is not explained by the variables included in the regression model.

A3.5 Coral Reefs

The average coral reef value is US$4,422 per hectare per year, and the median is US$772 per hectare per year. The explanatory variables included in the value function are as follows: the area of coral cover at the study site; the GDP per capita of the country in which the study site is located; the population within a 50 kilometer radius of the site; the length of roads within 50 kilometers; the human appropriation of net primary product (HANPP) within 50 kilometers; the net primary product within 50 kilometers; and the area of coral cover within 50 kilometers of the study site.

The value function is presented in Table A3.4. The estimated coefficients on the explanatory variables all have the expected signs, but only the area of coral cover at the study site is statistically significant. The negative coefficient on the area of coral cover indicates diminishing returns to scale. The positive effect of the income variable (GDP per capita) indicates that coral reef ecosystem services have higher values in countries with higher incomes; that is, coral reef ecosystem services are normal goods for which demand increases with income. This variable, however, is difficult to define and interpret clearly, since the beneficiaries of coral reef ecosystem services are often not from the country in which the reef is located. This is the case for most tourism and recreational services. The positive effect of population on the value of coral reef ecosystem services relates to market size or demand for ecosystem services. A larger population in the vicinity of a coral reef means that more people benefit from the ecosystem services that it provides. The negative effect of the length of roads in the vicinity of a coral reef reflects the associated effect of fragmentation and degradation on shore. Similarly, the negative effect of HANPP indicates the extent of human exploitation of natural resources in the region. The

Table A3.4 Coral reefs valuation function

Variable name	*Variable definition*	β	*SE*	*Sig.*
Constant		16.093	3.707	0.000
AREA_LN	Natural log of the study site area	–0.293	0.066	0.000
GCP50_IN	Natural log of Gross Cell Product within 50 kilometer radius	0.039	0.099	0.695
POP50_LN	Natural log of population within 50 kilometer radius of study site	0.238	0.154	0.125
RDS50_LN	Natural log of length of roads within 50 kilometer radius of study site	–0.035	0.107	0.743
HAN50_LN	Natural log of human appropriation of net natural product within 50 kilometer radius of study site	–0.076	0.054	0.161
NNP50_LN	Natural log of net natural product within 50 kilometer radius of study site	–0.379	0.287	0.189
COR50_LN	Natural log of area of coral reef within 50 kilometer radius of study site	–0.207	0.231	0.372
N		163		
Adjusted R^2		0.18		

Source: Hussain et al. (2011).

conversion of natural land uses or cultivation of crops often results in increased sedimentation in coastal waters, which can negatively affect reefs. The negative coefficient on the area of coral reefs in the vicinity of a specific reef indicates substitution effects between patches of coral reef. In areas where coral reefs are abundant, the value of a specific patch of reef will be lower.

The adjusted R^2 is relatively low (0.18), indicating that the estimated model explains only 18 percent of variation in coral reef values. There are clearly a number of important factors influencing the value of coral reefs that are not captured by this set of explanatory variables. The direction and magnitude of estimated effects of our set of explanatory variables do, nevertheless, follow theoretical expectations.

Notes

1. The review covered sources such as EconLit, EVRI database, and IUCN database for forest studies.
2. The discussion of the valuation functions for grasslands, wetlands, mangroves, and coral reefs is taken directly from Hussain et al. (2011).

Appendix 4

Description of the CGE Model

A4.1 The Economic Growth Baseline: GTAP-India Model

The following definitions apply to the model:

- Computable refers to numerically solvable models.
- General refers to an economy-wide approach.
- Equilibrium is satisfied at multiple levels among (1) demand and supply of factor of production, commodities, and services; (2) consumers' demands and their budget constraints (expenses equal revenues); and (3) macroeconomic balance [GDP = $C + G + I + (X - M)$].[1]

The GTAP model, like most of the standard CGE models, comprises nonlinear behavioral equations and macroeconomic accounting links (linear relations describing the break-even points in different markets).

The model is solved under GEMPACK (General Equilibrium Model Package), which uses a Euler algorithm and 3–4–5 step extrapolation method.

The Indian economy is modeled as an open economy composed of 57 firms, one representative household, and the government. Five factors of production exist: skilled labor, unskilled labor, capital, land, and natural resources.

Commodities and services, capital, and labor are mobile across sectors and countries (international migration is not specified in the current version). The model represents the circular flow of goods and services in the economy, which (1) permits flexibility in economic agents' behaviors, (2) captures substitution/complementarity relations across demand for goods and services, and (3) calculates

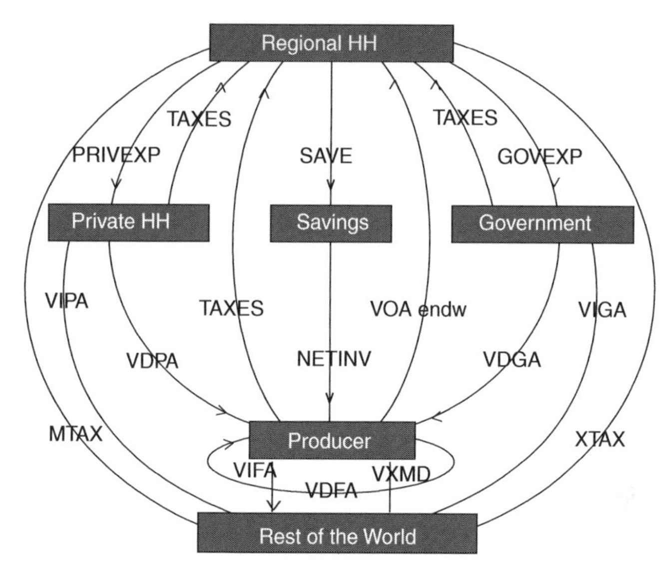

Figure A4.1 Circular flows in GTAP-CGE model

Adapted from http://www.unescap.org/tid/artnet/mtg/cb3_d3s2.pdf.

price changes resulting from changing demand and supply conditions (Figure A4.1).

Within a top-down structure, domestic gross output is an aggregate of domestic sales and exports obtained through a constant elasticity of transformation (CET) function. The production structure is specified in the form of nested constant elasticity of substitution (CES) functions that use labor (skilled and unskilled), capital, land, and natural resources as inputs (Figure A4.2).

Intermediate consumption include five energy products (coal, crude oil, petroleum products, natural gas, and electricity) and 52 nonenergy goods. All intermediate goods are differentiated according to their origin as domestic and imported products. Imports by the countries of origin follow an Armington specification (Armington, 1969).

Regional utility per capita is defined at the regional level, within a Cobb-Douglas function by private consumption, government consumption, and savings.

The demand for final goods is defined at the regional level by (1) household consumption through a constant-difference-elasticity (CDE)[2] demand specification, which is a nonhomothetic demand system, and (2) public sector using a Cobb-Douglas aggregation

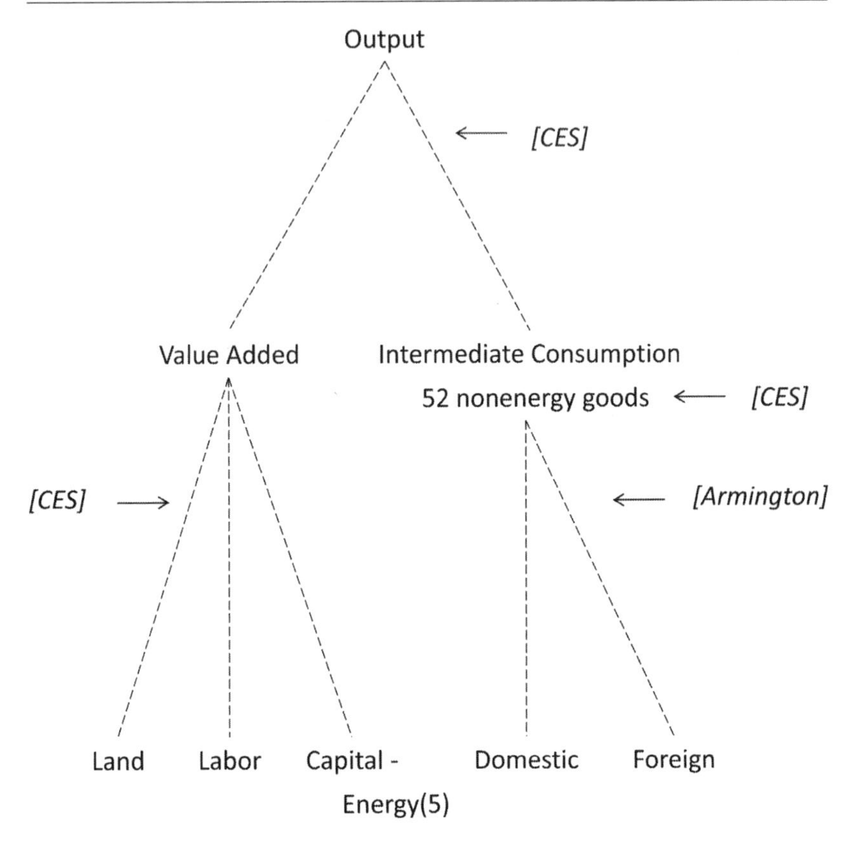

Figure A4.2 Modified production structure of the GTAP-CGE model used in this study

composed of market commodities and government spending where both are specified as a fixed share of income (Figure A4.3).

Household and firm savings, as well as taxes, finance investment and government expenses. The price of utility from private consumption depends on the level of private consumption expenditure.

The GTAP-India 2030 model is used to develop an economic baseline that represents the most likely path of development of the Indian economy until 2030. Population/labor force, capital inflows, and productivity growth are the drivers of the economic growth; no economic policy or pollution-control measures are specified.

The economic baseline is developed by applying shocks to the initial equilibrium conditions that represent the Indian economy and its linkages with the rest of the world in 2007.

In order to represent the most likely growth path, the model is solved for successive years using statistical projections on population,

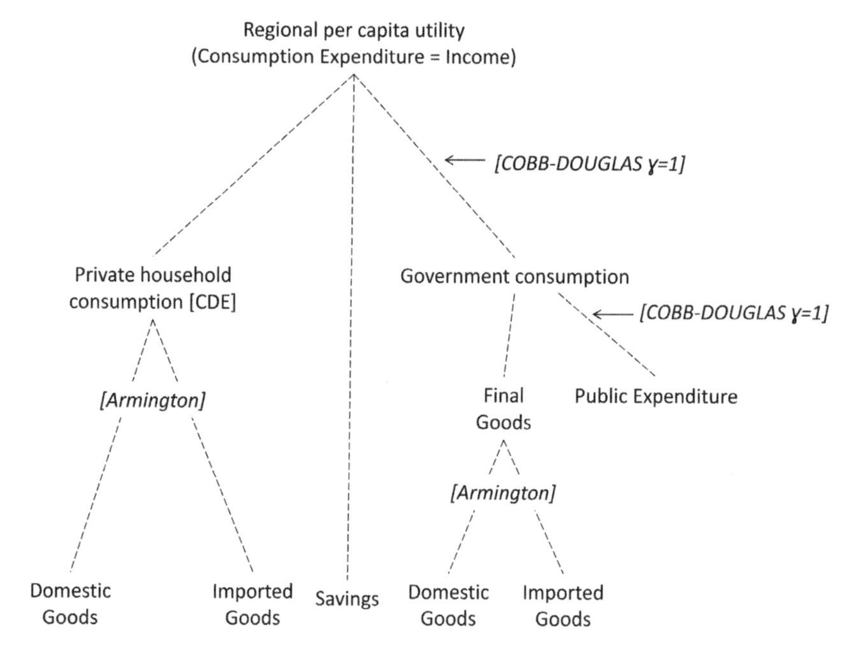

Figure A4.3 Consumption module in GTAP-CGE model

labor supply, and a total factor productivity of 2 percent per year based on the literature. A new equilibrium—that is, new prices and demand-and-supply conditions—are determined for each year.

A4.2 The PM10 Emission Baseline

Fossil fuels are the major source of many local and regional air pollutants, including suspended particulate matter (SPM) and PM10 emissions. Other sources of particulate matter include physical processes of grinding, crushing, and abrasion of surfaces. Mining and agricultural activities are also known to contribute larger-size particulate matter to the environment. In this exercise, the PM10 emissions, which cover the inhalable size fraction of SPM are estimated in two steps—as input- and output-related emissions.

The construction of the environmental baseline captures the influence of the economic growth drivers on India's SPM and PM10 levels. The emission estimates are introduced into the CGE model to calculate the economy-wide impacts of the various emission-reduction policies.

Equation 1 summarizes the PM10 estimation method: emissions (E) comprise input- and output-related pollutants. The former refers to fuel-combustion-related particulate emissions; therefore it is estimated on the basis of different categories of agents' demands for fuel (C). The second types of pollutants are emitted during the production processes (XP) of different sectors.

$$E = \sum_i \sum_j \alpha_{ij} C_{i,j} + \sum_i \beta_i XP_i \quad (1)$$

E = PM10 emissions
$C_{i,j}$ = Demand for energy products j
i = institution (firm, household, government)
j = energy product (coal, crude oil, natural gas, electricity, refined oil).
$\alpha_{i,j}$ = emission coefficient associated with the consumption of one unit of energy product j by the institution i
XP_i = Output of institution i
β_i = emission coefficient associated with one unit of output in sector i.

Most of the CO_2 emissions from fuel combustion are directly correlated to the level of carbon-intensive activities, such as electricity generation, production of chemicals and basic metal products, and consumption of transport fuels; these refer to direct production-based emissions.

This study borrows inputs from previous studies on India. More specifically, PM10 estimations developed by the International Institute for Applied System Analysis' (IIASA) GAINS model are used as the initial pollution level in our model. Accordingly, we assumed that the PM10 emission level corresponded to 7 million tons in 2005.

Based on the sectoral breakdown displayed in Appendix Table A4.1, we calculated the shares of PM10 emissions from fuel use in the GTAP model. Approximately three-quarters of PM10 emissions are caused by fuel consumption (5 million out of 7 million tons).

Table A4.1 shows that 93 percent of the energy consumption at the origin of the PM10 emissions was domestically produced in 2005. Carbon-intensive consumption accounts for 63 percent of the pollution.

PM10 emissions estimations linked to production process follow the method in the Garbaccio, Ho, and Jorgenson (2000) study for China. They are assumed to represent a certain percentage of the total PM10 emissions. The corresponding coefficient is borrowed

Table A4.1 PM10 estimations per fuel use (GAINS Model Simulations)

PM10 (thousand tons)	*1990*	*1995*	*2000*	*2005*
	5,702	6,731	7,059	7,032
Brown coal/lignite, grade 1	283	411	360	305
Hard coal, grade 2	1,485	2,023	2,049	2,041
Derived coal (coke, briquettes)	5	4	3	2
Biomass fuels	...	1	0	1
Agricultural residuals, direct use	608	696	802	764
Biogas	0	0	0	0
Dung	814	773	732	590
Fuelwood direct	938	1,020	1,097	1,228
Heavy fuel oil	6	7	8	9
Medium distillates (diesel, light fuel oil)	98	132	180	139
Gasoline and other light fractions of oil (includes kerosene)	83	117	162	67
Liquified petroleum gas	1	1	2	3
Natural gas (including other gases)	0	0	0	0
Non-exhaust PM emissions, road abrasion	3	3	4	3
Non-exhaust PM emissions, brake wear	1	2	2	1
Non-exhaust PM emissions, tire wear	3	4	5	3
No fuel use	1,373	1,356	1,651	1,876

Source: www.iiasa.ac.at

from the estimations developed by IIASA using the GAINS model. In 2005 output-related PM10 emission represented approximately 26 percent of the total PM10 level (Table A4.2 and A4.3)[3].

Currently, India's major cities have severe air pollution problems, with average ambient concentrations of pollutants far in

Table A4.2 PM10 emissions per fuel use and origin in GTAP-India 2030 model (2005)

	PM10		
	Domestic	*Imported*	*Total*
Coal	3,070,247	216,773	3,287,020
Crude oil	233	16	250
Natural gas	259,912	145	260,057
Refined oil and coal products	1,475,492	128,605	1,604,098
Electricity	0	0	0
Total	4,805,884	345,540	5,151,424

Table A4.3 GTAP mapping of process-based emission coefficients

	Process particulate emissions	
	GTAP	%
Agriculture	agr	0.00
Coal mining	coa	3.14
Crude petroleum	oil	3.14
Metal ore mining	omn	3.14
Other nonmetallic ore mining	omn	3.14
Food manufacturing	ofd	0.35
Textiles	tex	0.15
Apparel and leather products		0.00
Lumber and furniture manufacturing	lum	0.46
Paper, cultural, and educational articles	ppp	0.46
Electric power	ely	2.79
Petroleum refining	p_c	2.21
Chemicals	crp	2.75
Building material	nfm	57.76
Primary metals	I_s	12.27
Metal products	fmp	0.19
Machinery	ome	0.43
Transport equipment	mvh	0.43
Electric machinery andinstruments	ele	0.43
Electronic and communication equipment	ele	0.43
Instruments and meters	ome	0.43
Other industry	omf	5.92
Construction		0.00
Transportation and communications		0.00
Commerce		0.00
Public utilities		0.00
Culture, education, health, and research		0.00
Finance and insurance		0.00
Public administration		0.00
Households		0.00
Total		100.00

Source: Garbaccio et al. (2000).

excess of WHO guidelines and Indian ambient standards. These pollution levels and their adverse effects on human health are expected to increase with the rise of PM10 pollutants, which is detailed in the next section.

A4.3 Sectoral Sources of Particulate Emissions

Agriculture Related

Paddy or rice; wheat and other cereal grains; vegetables, fruits, nuts; oil seeds; sugar cane, sugar beet; plant-based fibers; cattle, sheep, and goats; others animal products; raw milk; wool and silk-worm cocoons; forestry; and fishing.

Energy Related

Coal mining; crude oil; natural gas extraction; refined oil products; petroleum; coal products; and electricity.

Energy-Intensive Industries

Minerals and mineral products; chemical, rubber, plastic products; ferrous metals; other metals.

Other Industries and Services

Meat products; vegetable oils and fats; dairy products; processed rice; sugar; food products; beverages; tobacco products; textiles and apparel; leather products; wood products; paper products, publishing; metal products; motor vehicles and parts; transport equipment; electronic equipment; machinery; manufacturing; water utilities; construction; manufacturing and distribution of natural gas; trade; water transport; air transport; communication; financial services; insurance; business services; recreational service; public administration and defene, education; ownership of dwellings.

A4.4 Assumptions of Business-as-Usual

Tables A4.4 and A4.5 give the key exogenous assumptions that were used in the model.

Table A4.4 Assumptions of Business-as-usual (Conventional GDP Growth) (%)

Year	*Population growth*	*Labor force growth*	*TFP change*	*GDP growth*
2010	1.01	1.01	2.00	10.08
2011	1.01	1.01	2.00	7.84
2012	1.01	1.01	2.00	7.51
2013	1.01	1.01	2.00	8.11
2014	1.01	1.01	2.00	8.17
2015	1.01	1.01	2.00	8.14
2016	1.01	1.01	2.00	8.16
2017	1.01	1.01	2.00	7.60
2018	1.01	1.01	2.00	7.23
2019	1.01	1.01	2.00	6.77
2020	1.01	1.01	2.00	6.58
2021	1.01	1.01	2.00	6.43
2022	1.01	1.01	2.00	6.24
2023	1.01	1.01	2.00	6.13
2024	1.01	1.01	2.00	5.95
2025	1.01	1.01	2.00	5.84
2026	1.01	1.01	2.00	5.77
2027	1.01	1.01	2.00	5.63
2028	1.01	1.01	2.00	5.57
2029	1.01	1.01	2.00	5.47
2030	1.01	1.01	2.00	5.37

A4.5 The Health Impact Simulations

Particulate matter can be defined as a mixture of liquid and solid particles and chemicals that vary in size and across space. The smaller the size of the particle, the easier it is for it to enter the human respiratory system and even the bloodstream in some cases. The existing literature on the health effects of particulate matter show that particles measuring less than 10 microns penetrate the lungs more easily than the larger-size particles. In particular, PM10 has an impact on respiratory diseases. Long-term exposure to PM10 can affect mortality and morbidity levels; the most widely known adverse health impact of PM10 is premature mortality. Levels of particulate matter are often much higher in developing countries as compared to those in developed countries. Ostro (1994) coefficients are used to calculate PM10 health effects.

Since PM10 causes premature death, one of the implications of high PM10 levels in the country would be a decrease in the available

Table A4.5 PM10 emission coefficients linked to fuel use by sector (tons per million, local currency)

Sector	*Coal*	*Oil*	*Natural gas*
Agriculture	42,560	160	27
Coal mining	38,182	143	24
Crude petroleum	38,182	143	24
Metal ore mining	38,182	143	24
Other nonmetallic ore mining	38,182	143	24
Food manufacturing	32,983	124	21
Textiles	18,505	69	12
Apparel and leather products	7,678	29	5
Lumber and furniture manufacturing	25,629	949	27
Paper, cultural and educational articles	25,629	949	27
Electric power	32,642	544	0
Petroleum refining	7,235	723	12
Chemicals	17,898	1,790	30
Building material	13,454	1,345	22
Primary metals	6,379	638	11
Metal products	8,814	33	6
Machinery	11,970	45	7
Transport equipment	11,970	45	7
Electric machinery and instruments	11,970	45	7
Electronic and communication equipment	11,970	45	7
Instruments and meters	11,970	45	7
Other industry	46,872	176	29
Construction	42,560	160	27
Transportation and communications	42,560	5,320	27
Commerce	42,560	160	27
Public utilities	42,560	160	27
Culture, education, health, and research	42,560	160	27
Finance and insurance	42,560	160	27
Public administration	42,560	160	27
Households	21,280	426	27

Source: Garbaccio et al., 2000.

labor force. With a large population, the effect on mortality rates and on the labor force are calculated using estimates from a Jakarta study (Ostro, 1994). The health damages calculated outside of the CGE model produced central, upper, and lower estimates for the coefficients of change in mortality. All three figures have been used to calculate a range of results in the CGE simulations.

As for the PM10 emissions, the projections from the CGE results have been used to calculate the PM10 concentrations for India. The base year concentration level of 97.58 $\mu g/m^3$ is the average concentration level of the pollutant across all cities in India (calculated using data from Central Pollution Control Board of the India environmental data bank). The concept of uniform rollback was used to calculate the concentrations for the subsequent years. Uniform rollback states that the percent change in pollutant emissions on an annual basis will equal the percent change in pollutant concentrations on an annual basis. Therefore, using the base year average for PM10 for the year 2010, projections can be made for PM10 concentrations using the percent change in PM10 emissions from the CGE model.

The dose–response coefficients are from the Ostro (1994) study on Jakarta. Such an epidemiological study has not been carried out for India. Jakarta is the next best study as its data provides more plausible health estimates than data from industrialized nations (Table A4.6). For the purpose of this study, we calculated the impact of premature mortality on India's labor force. Literature suggests that in general children and people above the age of 65 are most vulnerable to respiratory diseases from particulate matter In the case of India, however, the labor force will have maximum exposure to PM10 since they have maximum outdoor exposure.

The dose–response coefficient for premature mortality has a central value and upper and lower bounds for the 95-percent confidence interval. The numbers in Table A4.6 give the percentage increase in mortality from the baseline per one microgram per normal cubic meter of concentration. All three coefficients have been used to project a range of mortality effects along with the central estimates.

The total labor force numbers and projections have been obtained from the CGE model results. These numbers are used to calculate the affected labor force. Exposure to PM10 will reduce the labor force as a result of premature mortality. These numbers

Table A4.6 Dose–response coefficients

Dose–Response Coefficient	*Value*
Upper	0.008272
Central	0.006015
Lower	0.003758

Source: Pope et al. (1995).

will be used to project an economic growth path, taking into account the reduced labor force.

Health Impacts and Monetary Losses

The health damage estimates from PM10 were calculated for three health-related endpoints:

1. premature mortality from PM10
2. morbidity from PM10 in terms of reduced activity days (RAD)
3. morbidity from PM10 in terms of respiratory hospital admissions (RHA)

Premature Mortality

The log linear method has been used to estimate premature mortality, as outlined by WHO (2004). To estimate premature mortality, we used PM2.5 concentrations, which have been converted from PM10 using a conversion factor of 0.65. In order to calculate mortality, the relative risk (RR) is calculated based on the observed PM concentrations as shown in equation 2. Using the RR, the attributable factor (AF) is calculated as shown in equation 2. Premature mortality is estimated using equation 3 for all cities.

$$RR = [(X + 1)/(X_0 + 1)]^{\beta} \quad (2)$$

Where:
RR = relative risk
X = observed PM concentration
X_0 = background PM concentration (taken as 5 μg/m^3, as per WHO [2004] guidelines)
β = concentration-response coefficient

$$AF = (RR - 1)/RR \quad (3)$$

Where,
AF = attributable factor

$$\text{Mortality} = AF \times POP \times CMR \quad (4)$$

Where,
POP = city population exposed to PM2.5
CMR = urban crude mortality rate

PM2.5 is known to cause premature mortality, and the crude mortality rate (CMR) is required for its estimation. The CMR estimation was specifically done for urban areas (Office of the Registrar General and Census Commissioner of India, 2009). The CMR figure is higher at the national level than at the urban level, since the national CMR also includes deaths in rural areas. To obtain accurate results, urban CMR figures were used for mortality calculations. CMR projections were made following the trends in the past years since there is no other source for the CMR.

The dose–response coefficient for premature mortality as a result of exposure to PM2.5 was taken from Pope et al. (2002). Premature mortality estimates for the selected cities for all years were made using central estimates and the 95-percent confidence intervals for premature mortality.

The monetary value of premature mortality from PM2.5 was estimated using the standard Value of Statistical Life (VSL) method. VSL was estimated for premature deaths across the megacities, million-plus cities, and metropolitan cities from 2010 to 2030. This study used an average VSL based on estimates from four India-specific studies.[4] The values were as follows:

- Shanmugam (1997) using a WTP (willingness to pay) approach: Rs 18,932,020 (US$420,712)
- Simon et al. (1999) using a WTP approach: Rs 16,197,563 (US$359,946)
- Madheswaran (2007) using a compensating wage differential approach: Rs 16,939,353 (US$376,430)
- Bussolo and O'Connor (2001) using a human capital approach: Rs 19,109,280 (US$424,651)

The average VSL estimate from these four studies is US$404,422. This value will increase over time in line with the growth rate for income per capita as projected in the CGE model.

Reduced Activity Days (RADs)

The equation for calculating RADs due to PM10 exposure was as follows:

$$RAD = \gamma \times POP \times \mathrm{PM10}$$

Where,
RAD = reduced activity days from PM10 for a given year for each city
γ = RAD dose–response coefficient for PM10 (WHO, 2004)
POP = city population exposed to PM10
PM10 = PM10 concentration in each city

The WHO 2004 study estimated the dose–response coefficient for RAD arising from PM10 concentrations. The RAD coefficient was calculated based on epidemiological studies. The coefficient was used to determine RAD in each city until 2030. Reduced activity in a day would lead to a loss in income. The average income per capita per day in urban areas was used as the basis to determine the total loss. This income per capita per day increased in line with the projections for per capita GDP from the CGE model.

Respiratory Hospital Admissions (RHA)

The equation for calculating the respiratory hospital admissions from exposure to PM10 is as follows:

$$RHA = \xi \times POP \times \mathrm{PM10}$$

Where,
RHA = respiratory hospital admissions from PM10 for a given year for each city
ξ = RHA dose–response coefficient for PM10 (World Health Organization, 2004)
POP = city population exposed to PM10
PM10 = PM10 concentration in each city

The WHO 2004 study estimates the dose–response coefficient for RHA arising from PM10 concentrations. Each RHA involved an eight-day hospital stay, with incurred medical expenses and loss of income. The hospital costs were estimated at US$30 per day, based on WHO figures for India. The income per capita per day in urban areas was used as the basis to determine the total loss. Both the income per capita per day and hospital costs increased in line with the projections for per capita GDP from the CGE model.

Bibliography

Abbey, D. E., Lebowitz, M. D., Mills, P. K., Petersen, F. F., Beeson, W. L., Burchette, R. J. (1995). Long-term Ambient Concentrations of Particulates and Oxidants and Development of Chronic Disease in a Cohort of Nonsmoking California Residents. Inhalation Toxicology, 7: 19–34.

Ajai, A., Dhinwa, P., Pathan, S., and Ganesh Raj, K. (2009). Desertification/Land Degradation Status Mapping of India. *Current Science*, 97(10): 1478–1483.

Alberini, A., Loomes, G., Scasny M., and Bateman, I. (2010). *Valuation of Environment-Related Health Risks for Children*. Paris: Organization for Economic Co-Operation and Development.

Alkemade, R., Bakkenes, M., Bobbink, R., Miles, L., Nellemann, C., Simons, H., and Tekelenberg, T. (2002). GLOBIO 3: Framework for the assessment of global terrestrial biodiversity. In Bouwman, A., F., Kram, T., and Goldewijk, K. K.(eds.). Integrated Modeling of Global Environmental Change: An Overview of IMAGE 2.4. The Netherlands Environmental Assessment Agency, Bilthoven, pp. 171–186.

Alkemade, R., Van Oorschot, M., Miles, L., Nellemann, C., Bakkenes, M., and Ten Brink, B. (2009). GLOBIO3: A Framework to Investigate Options for Reducing Global Terrestrial Biodiversity Loss. *Ecosystems, 12*, 374–390.

Allen, R.G., Pereira, L.S., Raes, D., Smith, M. (1998). FAO Irrigation and Drainage Paper 56. Rome: FAO.

Armington, P. S. (1969). A Theory of Demand for Products Distinguished by Place of Production. *IMF Staff Papers*, 16 (pp. 159–178). Washington, D.C.: IMF.

Arrow, K. J., and Debreu, G. (1954). The Existence of an Equilibrium for a Competitive Economy. *Econometrica*, 22: 265–290.

Asian Development Bank. (2003). *Technical Assistance to India for Conservation and Livelihoods Improvement in the Indian Sundarbans* (TAR: IND 34272). Manila: ADB.

———. (2004). *Country Environmental Analysis*. Manila: ADB.

Asian Development Bank, Global Environment Facility, United Nations Development Programme (1998). *Asia Least Cost Greenhouse Gas Abatement Strategy: ALGAS India, Country Report*. Manila: ADB.

Barbier, E., Acreman, M., and Knowler, D. (1997) *Economic Valuation of Wetlands: A Guide for Policy Makers and Planners*. Gland, Switzerland: Ramsar Convention Bureau. Available at: http://cmsdata.iucn.org/downloads/03e_economic_valuation_of_wetlands.pdf.

Beghin, J., Bowland, B., Dessus, S., Roland-Holst, D., and Van der Mensbrugghe, D. (2002). Trade Integration, Environmental Degradation and Public Health in Chile: Assessing the Linkages. *Environment and Development Economics*, 7(2): 241–267.

Beghin, J., Dessus S., Roland-Holst, D., and Van der Mensbrugghe, D. (1996). *General Equilibrium Modelling of Trade and the Environment* (Technical Paper, N8 116). Paris: OECD Development Center.

Bergman, L. (1991). General Equilibrium Effect of Environmental Policy: A CGE Modeling Approach. *Environmental and Resource Economics*, 1: 43–61.

Bhalla, S. (2007). *Second Among Equals: The Middle Class Kingdoms of India and China*. Washington, D.C.: Institute of International Economics.

Bose, S. (2002). The Sundarbans Biosphere: A Study on Uncertainties and Impacts in Active Delta Region. Centre for Built Environment, Kolkata. *Proceedings of the 2nd Asia Pacific Association of Hydrology and Water Resources (APHW) Conference* (pp. 475–483). Singapore.

———. (2007). Mangrove Forests in Sunderbans Active Delta: Ecological Disaster and Remedies. 10th International River Symposium and Environmental Flows Conference, River Symposium and the Nature Conservancy, Brisbane, Australia.

Bovebourg, A. L., and Goulder, L. H. (1996). Optimal Environmental Taxation in the Presence of Other Taxes: General-Equilibrium Analyses. *American Economic Review*, 86(4): 985–1000.

Brandon, C., and Homman, K. (1995). *The Cost of Inaction: Valuing the Economy-wide Cost of Environmental Degradation in India*. Asia Environmental Division, Washington, D.C.: World Bank.

Brock, W. A., and Xepapadeas, A. (2003). Valuing Biodiversity from an Economic Perspective: A Unified Economic, Ecological, and Genetic Approach. *The American Economic Review*, 93(5): 1597–1614.

Bruner, A., Gullison, R. E., and Balmford, A. (2004). Financial Needs for Comprehensive, Functional Protected Area Systems in Developing Countries. *BioScience*, 54: 1119–1126.

Burniaux, J. M., and Truong, T. P. (2002). *GTAP-E: An Energy Environmental Version of the GTAP Model with Emissions* (Technical Paper 16). West Lafayette, IN: Centre for Global Trade Analysis.

Bussolo, M., and O'Connor, D. (2001). *Clearing the Air in India: The Economics of Climate Policy with Ancilliary Benefits* (Development Centre Working Paper 182, CD/DOC 14). Paris: OECD.

Campbell-Lendrum, D., and Woodruff, R. (2007). *Climate Change: Quantifying the Health Impact at National and Local Levels* (Environmental Burden of Disease Series, no. 14). Geneva: Public Health and the Environment.

Central Pollution Control Board. (2006). *Air Quality Trends and Action Plan for Control of Air Pollution from Seventeen Cities*. Available at: http://www.cpcb.nic.in/upload/NewItems/NewItem_104_airquality 17cities-package-.pdf.

———. (2009a). *Ambient Air Quality Standards*. Available at: http://cpcbenvis.nic.in/airpollution/standard.htm.

———. (2009b). *Comprehensive Environmental Assessment of Industrial Clusters*. Available at: http://www.cpcb.nic.in/upload/NewItems/NewItem_152_Final-Book_2.pdf.

———. (2010). *National Air Quality Monitoring Program*. Available at: http://cpcbenvis.nic.in/airpollution/finding.htm.

———. (2011). *Air Quality Monitoring, Emission Inventory and Source Apportionment Study for Indian Cities: National Summary Report*. Available at: http://cpcb.nic.in/FinalNationalSummary.pdf.

Central Pollution Control Board and the Environmental Data Bank. (2009). *RSPM Emission Data by City, March 2009 to June 2009*. Available at: www.cpcbedb.nic.in.

Central Soil Salinity Research Institute. (2010). *Annual Report 2010–2011*. Karnal, India: CSSRI.

Chiabai, A., Travisi, C., Markandya, A., Ding, H., and Nunes, P.A.L.D. (2011). Economic Assessment of Forest Ecosystem Services Losses: Cost of Policy Inaction. *Environmental and Resource Economics*. doi: 10.1007/s10640–011–9478–6.

Chinese Academy for Environmental Planning (CAEP) and The Energy and Resources Institute (TERI). (2010). *Environment and Development: China and India*. New Delhi: TERI Press.

Cohen, A. et al. (2004). Urban Air Pollution. In M. Ezzati et al. (Ed.), *Comparative Quantification of Health Risks*. Geneva: WHO.

Conrad, K., and Henseler-Unger, I. (1986). Applied General Equilibrium Modeling for Longterm Energy Policy in the Fed. Rep. of Germany. *Journal of Policy Modeling*, 8(4): 531–549.

Croitoru, L., and Sarraf, M. (2010). *The Cost of Environmental Degradation: Case Studies from the Middle East and North Africa* (Directions in Development series). Washington, D.C.: World Bank.

Cropper, M., Gamkhar, S., Malik, K., Limonov, A., and Partridge, I. (2012). *The Health Effects of Coal Electricity Generation in India, Resources for the Future* (Discussion Paper, RFF DP 12–25). Washington, D.C.: RFF.

Cropper, M., Nathalie, B. S., Anna, A., Seema, A., and Sharma, P. K. (1997). The Health Benefits of Air Pollution Control in Delhi. *American Journal of Agricultural Economics*, 79(5): 1625–1629.

Cropper, M., and Oates, W. (1992). Environmental Economics: A Survey. *Journal of Economic Literature*, 30: 675–740.

Cullu, M. A. (2003). Estimation of the Effect of Soil Salinity on Crop Yield Using Remote Sensing and Geographic Information System. *Turkish Journal of Agriculture and Forestry*, 27(1): 23–28.

Dasgupta, P. (2011). *Inclusive National Accounts: Introduction* (Working paper prepared for the Expert Group for Green National Accounting for India). New Delhi, India: Ministry of Statistics and Programme Implementation (MOS&PI), Government of India.

Desai, M., Mehta, S., and Smith, K. (2004). *Indoor Smoke from Solid Fuels. Assessing the Environmental Burden of Disease at National and Local Levels* (Environmental Burden of Disease Series, no. 4). Geneva: WHO.

Dockery, D. W., Pope, C. A. III, Xu, X., et al. (1993). An Association between Air Pollution and Mortality in Six US Cities. *New England Journal of Medicine*, 329: 1753–1759.

Downing, M., and Ozuna, J. (1996). Testing the Reliability of the Benefit Function Transfer Approach. *Journal of Environmental Economics and Management*, 30: 316–322.

Dufournaud, M., Harrington, J., and Rogers, P. (1988). Leontief's Environmental Repercussions and the Economic Structure Revisited: A General Equilibrium Formulation. *Geographical Analysis*, 20(4): 318–327.

Edwards, T. H., and Hutton, J. P. (2001). Allocation of Carbon Permits within a Country: A General Equilibrium Analysis of the United Kingdom. *Energy Economics*, 23(4): 371–386.

EM-DAT-International Disaster Database. (2011). *Annual Disaster Statistical Review 2011*. Geneva: Center for Research on Epidemiology of Disasters (CRED).

Esrey, S. A., and Habicht, J-P. (1988). Maternal Literacy Modifies the Effect of Toilets and Piped Water on Infant Survival in Malaysia. *American Journal of Epidemiology*, 127(5): 1079–1087.

Esrey S. A., Potash J. B., Roberts, L., and Shiff, C. (1991). Effects of Improved Water Supply and Sanitation on Ascariasis, Diarrhoea, Dracunculiasis, Hookworm Infection, Schistosomiasis, and Trachoma. *Bulletin of the World Health Organization*, 69(5): 609–621.

Fewtrell, L., and Colford, J. Jr. (2004). *Water, Sanitation and Hygiene: Interventions and Diarroea: A Systematic Review and Meta-analysis* (HNP Discussion Paper). Washington, D.C.: World Bank.

Fewtrell, F., Prüss, A., and Kaufmann, R. (2003). *Guide for Assessment of EBD at National and Local Level: Lead.* Geneva: WHO.

Food and Agriculture Organization. (2005). *Global Forest Assessment Report: India*. Rome: FAO.

———. (2011). *FAOSTAT Report*. Rome: FAO.

———. (2010). *Global Forest Resources Assessment: Main Report*. Rome: FAO.

Forest Survey of India. (2009). *India State of Forest Report*. New Delhi: Ministry of Environment and Forests, Government of India.

———. (2011). *India State of Forest Report*. New Delhi: Ministry of Environment and Forests, Government of India.

Forsund, F., and Storm, S. (1988). *Environmental Economics and Management: Pollution and Natural Resources*. New York: Croom Helm Press.

Garbaccio, R. F., Ho, M. S., and Jorgenson, D. W. (2000). The Health Benefits of Carbon Control in China. In *Proceedings of Workshop on Assessing the Ancillary Benefits and Costs of Greenhouse Gas Mitigation Strategies* (pp. 27–29), March. Washington, D.C.: Organization for Economic Co-operation and Development.

Garg, A. (2011). Pro-equity Effects of Ancillary Benefits of Climate Change Policies: A Case Study of Human Health Impacts of Outdoor Air Pollution in New Delhi. *World Development*, 39: 6.

Gopal, B., and Chauhan, M. (2006). Biodiversity and Its Conservation in the Sundarban Mangrove Ecosystem. *Aquatic Sciences*, 68(3): 338–354.

Gosain, A. K, Rao, S., and Busaray, D. (2006). Climate Change Impact Assessment on Hydrology of Indian River Basins. *Current Science*, 90(3): 346–353.

Goulder, L. (Ed.). (2002). *Environmental Policy Making in Economics with Prior Tax Distortions*. Northampton, MA: Edward Elgar.

Government of India. (2004). *India National Report on Implementation of United Nations Convention to Combat Desertification* (Submitted to United Nations Convention to Combat Desertification (UNCCD) Secretariat). Bonn, Germany.

Government of India, Ministry of Statistics and Program Implementation. (2011). *India Statistical Yearbook 2010*. New Delhi: Government of India, Ministry of Statistics and Program Implementation.

Gratten, Z., and Roberts, S. (2002). Rice Is More Sensitive to Salinity than Previously Thought. *California Agriculture*, 56(6), 189–195.

Greenstone, M., Krishnan, A., Ryan, N., and Sudarshan, A. (2012). Improving Human Health Through a Market-Friendly Emissions Scheme. *Seminar Volume for International Seminar on Global Environment and Disaster Management: Law and Society*. New Delhi: Supreme Court of India, Ministry of Environment and Forest, and Law Ministry.

Grossman, G., and Krueger, A. (1993). Environmental Impacts of a North American Free Trade Agreement. In P. Garber (Ed.), *The Mexico–US Free Trade Agreement*. Cambridge, MA: MIT Press.

Gundimeda, H. (2000). Estimation of Biomass in Indian Forests. *Biomass and Bioenergy*, 19(4): 245–258.

———. (2001). Managing Forests to Sequester Carbon. *Journal of Environmental Planning and Management*, 44(5): 701–720.

Gundimeda, H., Sanyal, S., Sinha, R., and Sukhdev, P. (2005). *The Value of Timber, Carbon, Fuelwood, and Non-Timber Forest Products in India's Forests* (Monograph 1: Green Accounting for Indian States Project). New Delhi:TERI Press.

Gundimeda, H., Sanyal, S., Sinha, R., and Sukhdev, P. (2006). *The Value of Biodiversity in India's Forests* (Monograph 4: Green Accounting for Indian States and Union Territories Project). New Delhi: TERI Press.

Gundimeda, H., and Sukhdev, P. (2008). GDP of the Poor. In P. ten Brink (Ed.), *The Economics of Ecosystems and Biodiversity in National and International Policy Making*. London: Earthscan.

Gunning, J. W., and Keyzer, M. (1995). Applied General Equilibrium Models for Policy Analysis. In J. Behrman and T. N. Srinivasan (Eds.), *Handbook of Development Economics Vol. III-A* (pp. 2025–2107). Amsterdam: Elsevier.

Guttikunda, S., and Jawahar, P. (2011). *Urban Air Pollution and Co-benefits Analysis in India*. New Delhi: Climate Works Foundation.

Hadker, N., Sharma, S., David, A., and Muraleedharan, T. R. (1997). Willingness-to-Pay for Borivli National Park: Evidence from a Contingent Valuation. *Ecological Economics*, 21: 105–122.

Haines-Young, R., and Potschin, M. (2010). The Links between Biodiversity, Ecosystem Services and Human Well-being. In D. Raffaelli and C. Frid (Eds.), *Ecosystem Ecology: A New Synthesis*. Cambridge: Cambridge University Press.

Harrison, G.W., Rutherford, T. F., and Tarr, D.G. (1997). Quantifying the Uruguay Round. *Economic Journal*, 107: 1405–1430.

Hazilla, M., and Koop, R. (1990). Social Cost of Environmental Quality Regulations: A General Equilibrium Analysis. *Journal of Policy Modeling*, 98(4): 853–873.

Hertel, T. et al. (1997). *Global Trade Analysis Modeling and Applications*. New York: Cambridge University Press.

Hocking, D., and Mattick, A. (1993). Dynamic Carrying Capacity Analysis as Tool for Conceptualising and Planning Range Management Improvements, with a Case Study from India (Vol. 34, Pt. 3). London: Overseas Development Institute.

Hussain, S. et al. (2011). *The Economics of Ecosystems and Biodiversity: Quantitative Assessment* (Draft final report to the United Nations Environment Programme). Available at: http://www.fao.org/nr/water/aquastat/countries/india/index.stm.

Ianchovichina, E., Roy, D., and Shoemaker, R. (2001). Resource Use and Technological Progress in Agriculture: A Dynamic General Equilibrium Analysis. *Ecological Economics*, 38: 275–291.

Indian Council of Forestry Research and Education. (2001). *Forestry Statistics of India 1987–2001*. Dehradun: ICFRE.

Indian Grassland and Fodder Research Institute. (2010). *Annual Report 2009–10*. Jhansi, India: IGFRI.

International Monetary Fund. (2011). *World Economic Outlook: Slowing Growth, Rising Risks*. Washington, D.C.: IMF.

———. (2012). *India: 2012 Article IV Consultation: Staff Report*. Washington, D.C.: IMF.

Jethoo, A. S., and Poonia, M. P. (2011). Water Consumption Pattern of Jaipur City (India). *International Journal of Enviromental Science and Development*, 2(2): 152–155.

Jorgenson, D. W., and Wilcoxen, P. (1990). Intertemporal General Equilibrium Modeling of U.S. Environmental Regulation. *Journal of Policy Modeling*, 12(4): 715–744.

———. (1993). Reducing U.S. Carbon Dioxide Emissions: An Assessment of Different Instruments. *Journal of Policy Modeling*, 15(5): 491–520.

Kadekodi, G. K., and Ravindranath, N. H. (1997). Macro-economic Analysis of Forestry Options on Carbon Sequestration in India. *Ecological Economics*, 23: 201–223.

Kapur, D., Ravindranath, D., Kishore, K., Sandeep, K., Priyadarshini, P., Kavoori, P. S., Chaturvedi, R. and Sinha, S. (2010). A Commons Story. In the Rain Shadow of Green Revolution. Anand (Gujarat), India: Foundation for Ecological Security.

Kirchhoff, S. (1998). Benefit Function Transfer vs. Meta-Analysis as Policy-Making Tools: A Comparison. In *Proceedings of Workshop on Meta-Analysis and Benefit Transfer: State-of-the-Art and Prospects*. Tinbergen Institute, Amsterdam, April 6–7.

Kishwan, J. et al. (2009). *India's Forest and Tree Cover: Contribution as a Carbon Sink* (Technical Paper no. 130). Dehradun, India: Indian Council of Forestry Research and Education.

Kotuby-Amacher, J. et al. (1997). *Salinity and Plant Tolerance*. Utah: Utah State University Press.

Krupnick, A., Larsen, B., and Strukova, E. (2006). *Cost of Environmental Degradation in Pakistan: An Analysis of Physical and Monetary Losses in Environmental Health and Natural Resources*. Washington, D.C.: World Bank.

Kumar, P., Sanyal, S., Sinha, R., and Sukhdev, P. (2006). *Accounting for the Ecological Services of India's Forests: Soil Conservation, Water Augmentation, and Flood Prevention* (Monograph 7: Green Accounting for Indian States and Union Territories Project). New Delhi: TERI Press.

Kumar, S., and Rao, D. N. (2003). Estimating Marginal Abatement Costs of SPM: An Application to the Thermal Power Sector in India. *Energy Studies Review*, 11(1): Article 2. Available at: http://digitalcommons.mcmaster.ca/esr/vol11/iss1/2.

Lampietti, J., and Dixon, J. (1994). *To See the Forest for the Trees: A Guide to Non-Timber Forest Benefits.* Washington, D.C.: World Bank.

Larsen, B. (2003). Hygiene and Health in Developing Countries: Defining Priorities through Cost-Benefit Assessments. *International Journal of Environmental Health Research*, 13: S37–S46.

———. (2004a). *Cost of Environmental Degradation: A Socio-Economic and Environmental Health Assessment in Damietta, Egypt* (Report prepared for Support for Environmental Assessment and Management [SEAM] programme implemented by the Egyptian Environmental Affairs Agency, Entec UK Ltd and ERM with support from the UK Department for International Development). Cairo, Egypt.

———. (2004b). *Cost of Environmental Damage: A Socio-Economic and Environmental Health Risk Assessment* (Final Report, prepared for the Ministry of Environment, Housing and Land Development, Republic of Colombia).

Lookman, A., and Rubin, E. S. (1998). Barriers to Adopting Least-Cost Particulate Control Strategies for Indian Power Plants. *Energy Policy*, 26(14): 1053–1063.

Louisiana Department of Public Health. (2004). *Typhoid Fever.* Available at: www.oph.dhh.state.la.us/infectiousdisease/docs/disease_updates_04/TyphoidManual.pdf.

Lucas, R., Wheeler, D., and Hettige, H. (1992). Economic Development, Environmental Regulation and the International Migration of Toxic Industrial Pollution: 1960–1988. In P. Low (Ed.), *International Trade and the Environment* (World Bank Discussion Paper 159, pp. 67–88). Washington, D.C.: World Bank

Ma, Q. (1999). *Asia-Pacific Forestry Sector Outlook Study: Volume I—Socio-Economic, Resources and Non-Wood Products* (Working Paper no. APFSOS/WP/43). Rome: Forestry Policy and Planning Division.

Mace, G. M., Gittleman, J. L., and Purvis, A. (2003). Preserving the Tree of Life. *Science*, 300: 1707–1709.

Madheswaran, S. (2007). Measuring the Value of Statistical Life: Estimating Compensating Wage Differentials among Workers in India. *Social Indicators Research*, 84(1): 83–96.

Madrid-Aris, M. E. (1998). International Trade and the Environment: Evidence from the North America Free Trade Agreement (NAFTA). Presented at the World Congress of Environmental and Resources Economics, Venice, Italy, June 25–27.

Mahal, A., Karan, A., and Engelgau, M. (2009). *The Economic Implications of Non-Communicable Diseases for India* (HNP Discussion Paper). Washington, D.C.: The World Bank.

Mani, M., Markandya, A., Sagar, A. and Strukova, E. (2012), An Analysis of Physical and Monetary Losses of Environmental Health and Natural Resources in India, Policy Research Working Paper No. 6219, Washington, DC: World Bank.

Markandya, A. (2006). Water Quality Issues in Developing Countries. In R. Lopez and M. A. Toman (Eds.), *Economic Development and Environmental Sustainability.* New York: Columbia University Press.

Martin,W., and Winters, L. A. (Eds.). (1996). *The Uruguay Round and the Developing Economies.* New York: Cambridge University Press.

Massachusette Institute of Technology. (2004). *Computable General Equilibrium Models and Their Use in Economy-Wide Policy Analysis* (Technical Note no. 6, September, 2004). Boston: MIT University Press.

McCann, K. S. (2000). The Diversity-stability Debate. *Nature*, 405: 228–233.

McDougall, R. (2007). GTAP-E 6 Pre-2. Available at: https://www.gtap.agecon.purdue.edu/resources/res_display.asp?RecordID = 2957.

Millennium Ecosystem Assessment (MEA). (2005). *Ecosystems and Human Wellbeing: Current State and Trends.* Washington, D.C.: Island Press.

Ministry of Environment and Forests. (2003). *Third National Report on Implementation of UN Convention to Combat Desertification* (Submitted to United Nations Convention to Combat Desertification [UNCCD] Secretariat, Bonn, Germany). New Delhi: Ministry of Environment and Forests.

Ministry of Food, Agriculture and Livestock (2006). Agricultural Statistics of India, Economic Wing, Ministry of Food, Agriculture and Livestock (2003–2004). New Delhi: Government of India.

———. (2006). *Report of the National Forest Commission.* New Delhi: Ministry of Environment and Forests, Government of India.

———. (2009a). *India's GHG Emissions Profile* (Climate Modelling Forum, supported by Ministry of Environment and Forests, September 2009). New Delhi: Ministry of Environment and Forests.

———. (2009b). *State of Environment Report.* Available at: http://envfor.nic.in/soer/2009/SoE%20Report_2009.pdf.

———. (2009c). *Asia-Pacific Forestry Sector Outlook Study II: India Country Report* (Working Paper no. APFSOS II/WP/2009/06). Bangkok: FAO.

———. (2009d). *India's Forest and Tree Cover: Contribution as a Carbon Sink.* New Delhi: Ministry of Environment and Forests, Government of India.

———. (2010). *India: Greenhouse Gas Emissions: Indian Network of Climate Change Assessment*. New Delhi: Ministry of Environment and Forests.

———. (2011). *Forest Survey of India 2011: India State of Forest Report*. New Delhi: Ministry of Environment and Forests, Government of India.

Ministry of Environment and Forests and GIZ. (2012). *The Economics of Ecosystems and Biodiversity—India Initial Assessment and Scoping Report* (Working document). New Delhi: Ministry of Environment and Forests, Government of India.

Ministry of Environment and Forests and World Bank. (2001). *The State of Environment Report 2001*. New Delhi: Ministry of Environment and Forests.

Ministry of Environment and Forests and World Bank. (2006). *India. Unlocking Opportunities for Forest-Dependent People in India*. Washington, D.C.: World Bank.

Mirza, S., Thomson, E., Akhbar, G., Sattar Alvi, A., and Rafique, S. (1996). *Balochistan: Searching for a Strategy*. Available at: http://www.icarda.org/Publications/Caravan/Caravan4/Car411.Html.

Mrozek, J., and Taylor, L. (2002). What Determines the Value of Life? A Meta Analysis. *Journal of Policy Analysis and Management*, 21(2): 253–270.

Murthy, N. S., Panda, M., and Parikh, J. (2000). *CO2 Emission Reduction Strategies and Economic Development in India* (IGIDR Discussion Paper). Mumbai: IGIDR.

Murty, M. N., Kumar, S., and Dhavala, K. (2006). Measuring Environmental Efficiency of Industry: A Case Study of Thermal Power Generation. In A. Lopez et al. (Eds.), *Global Burden of Disease and Risk Factors*. Washington, D.C.: World Bank. Available at: http://www.dcp2.org/pubs/GBD.

National Family Health Survey (NFHS-3) (2007). *India*. Mumbai: International Institute for Population Sciences.

National Institute of Rural Development. (2003). *Rural Development Statistics, 2002–03*. Hyderabad: National Institute of Rural Development.

National Sample Survey Organization. (2004). *Morbidity, Health Care and the Condition of the Aged* (Report No. 507 (60/25.0/1)). New Delhi: Government of India.

———. (2010). *Some Characteristics of Urban Slums July 2008–June 2009* (NSS 65th round, May 2010). New Delhi: Ministry of Statistics and Programme Implementation, Government of India.

Nayak, B.P., Kohli, P., and Sharma, J.V. (2012). Livelihood of Local Communities and Forest Degradation in India: Issues for REDD+ (Jointly prepared by Ministry of Environment and Forests and Tata Energy Research Institute). New Delhi:TERI Press.

Niederman, M. et al. (1999). Treatment Cost of Acute Exacerbations of Chronic Bronchitis. *Clinical Therapy*, 21(3): 576–591.

Nkonya, E., Gerber, N., von Braun, J., and De Pinto, A.(2011). *Economics of Land Degradation. The Costs of Action versus Inaction* (International Food Policy Research Institute Policy Brief N 68). Washington D.C.: International Food Policy Research Institute.

Office of the Registrar General and Census Commissioner of India. (2006). Population Projections for India and the States 2001–2026 (Revised December 2006). New Delhi:Census Commissioner of India.

———. (2009). *Sample Registration System Bulletin* (Vital Statistics Division, October). New Delhi: Census Commissioner.

Office of the Registrar General and Commissioner of Census. (2004). *Report on Causes of Death: 2001–03*. New Delhi.

Ojha, V. P. (2005). *The Trade-off among Carbon Emissions, Economic Growth and Poverty Reduction in India* (SANDEE Working Paper no. 12–05). Nepal: South Asian Network for Development and Environmental Economics.

Ojha, V. P. (2008). Carbon Emissions Reduction Strategies and Poverty Alleviation in India. *Environment and Development Economics,* 14: 323–348

Ostro, B. (1994). *Estimating the Health Effects of Air Pollution: A Method with an Application to Jakarta* (Policy Research Working Paper). Washington, D.C.: World Bank.

———. (2004). *Outdoor Air Pollution: Assessing the Environmental Burden of Disease at National and Local Levels* (WHO Environmental Burden of Disease Series, no. 5). Geneva: World Health Organization.

Ostro, B. & Chestnut, L. (1998). 'Assessing the Health Benefits of Reducing Particulate Matter Air Pollution in the United States'. ENVIRONMENTAL RESEARCH, 76: 94–106.

Othman, J. (2002). *Household Preferences for Solid Waste Management in Malaysia* (Economy and Environment Program for Southeast Asia(EEPSEA) Research Report no. 2002–RR8). Ottawa, Canada: International Development Research Centre.

Panagariya, A. (2008). *India: The Emerging Giant*. New York: Oxford University Press.

Parikh, K. (2009). *India's Energy Needs, CO_2 Emissions and Low Carbon Options*. Paper presented at the 22nd International Conference on Efficiency, Cost, Optimization, and Environmental Impact of Energy Systems, Foz do Iguaçu, Paraná, Brazil, August 31–September 3.

PBL. (2010). *Rethinking Global Biodiversity Strategies: Exploring Structural Changes in Production and Consumption to Reduce Biodiversity Loss*. The Hague: Netherlands Environmental Agency. Available at: www.pbl.nl/en/publications/2010/Rethinking_Global_Biodiversity_Strategies.html.

Pearce, D., Putz, F., and Vanclay, J. (1999). *A Sustainable Forest Future?* (CSERGE Working Paper GEC 99–15). London: CSERGE.

Perry, G., Whalley, J., and McMahon, G. (Eds.). (2001). *Fiscal Reform and Structural Change in Developing Countries*. New York: Palgrave-Macmillan.

Pigott, J., Whalley, J., and Wigle, J. (1992). International Linkages and Carbon Reduction Initiatives. In K. Anderson and R. Blackhurst (Eds.), *The Greening of World Trade Issues*. Ann Arbor: University of Michigan Press.

Pigou, A. (1932). *The Economics of Welfare*. London: MacMillan.

Planning Commission, Government of India. (2010). *Mid Term Appraisal Report: XIth Five Year Plan*. New Delhi: Planning Commission of India.

———. (2011). *Low Carbon Strategies for Inclusive Growth: An Interim Report*. New Delhi: Planning Commission of India.

Pope, C. A. III, Burnett, R. T., Thun, M. J., et al. (2002). Lung Cancer, Cardiopulmonary Mortality, and Long-Term Exposure to Fine Particulate Air Pollution. *Journal of the American Medical Association*, 287: 1132–1141.

Pope, C. A. III, Thun, M. J., Nambudiri, M. M., et al. (1995). Particulate Air Pollution as a Predictor of Mortality in a Prospective Study of US Adults. *American Journal of Respiratory and Critical Care Medicine*, 151: 669–674.

Prüss-Üstün, A., Fewtrell, F., and Landrigan P. (2004). Lead Exposure. In M. Ezzati et al. (Eds.), *Comparative Quantification of Health Risks: Global and Regional Burden of Disease Attributable to Selected Major Risk Factors*. Geneva: World Health Organization.

Prüss-Üstün, A., Kay, D., Fewtrell, L., and Bartram, J. (2004). Unsafe Water, Sanitation and Hygiene. In M. Ezzati et al. (Eds.), *Comparative Quantification of Health Risks: Global and Regional Burden of Disease Attributable to Selected Major Risk Factors*. Geneva: World Health Organization.

Purvis, A., and Hector, A. (2000). Getting the Measure of Biodiversity. *Nature*, 405(May 11): 212–219.

Ravindranath, N. H., Srivastava, N., Murthy, I. K., Malaviya, S., Munsi, M., and Sharma, N. (2012). Deforestation and Forest Degradation in India: Implications of REDD+. *Current Science*, 102(8): 1–9.

Ready, R., and Navrud, S. (2006). International Benefits Transfer: Methods and Validity Tests. *Ecological Economics*, 60: 429–434.

Robinson, S., and Gelhar, C. (1995). *Land, Water and Agriculture in Egypt: The Economy-Wide Impact of Policy Reform* (Discussion paper). Washington, D.C.: International Food Policy Research Institute.

Rodrik, D., and Subramanian, A. (2004). Why India Can Grow at 7% a Year or More. *Economic and Political Weekly* (Special Article, April 17): 1591–1596.

Rosenberger, R. S., and Johnston, R. J. (2009). Selection Effects in Meta-Analysis and Benefit Transfer: Avoiding Unintended Consequences. *Land Economics*, 85: 410–428.

Rosenberger, R. S., and Stanley, T. D. (2006). Measurement, Generalization, and Publication: Sources of Error in Benefit Transfers and Their Management. *Ecological Economics*, 60: 372–378.

Roy, A., Hu, H., Bellinger, D., Palaniapan, K., Wright, R., Schwartz, J., and Balakrishnan, K. (2009). Predictors of Blood Lead in Children in Chennai, India. *International Journal of Occupational and Environmental Health*, 15(4): 351–359.

Roy, D., and Roy, M. (1996). Communal Grazing Lands and Their Importance in India and Some Other Asian Countries. In *Proceedings of XVIII International Grassland Congress*, vol. 3. Calgary, Alberta: International Grassland Congress.

Roy, M. M., and Singh, K. A. (2008). The Fodder Situation in Rural India: Future Outlook. International Forestry Review, 10(2): 217–234.

Ruitenbeek, H. J. (1992). *Mangrove Management: An Economic Analysis of Management of Options with a Focus on Bintuni Bay, Irian Jaya.* Halifax, Canada: Environmental Management Development in Indonesia Project.

Ruitenbeek, H. J. (1994). Modelling Economy-Ecology Linkages in Mangroves: Economic Evidence for Promoting Conservation in Bintuni Bay, Indonesia. *Ecological Economics*, 10: 233–247.

Rutstein, S. O. (2000). Factors Associated with Trends in Infant and Child Mortality in Developing Countries during the 1990s. *Bulletin of the World Health Organization*, 78(10): 1256–1270.

Salkever, D. S. (1995). Updated Estimates of Earnings Benefits from Reduced Exposure of Children to Environmental Lead. *Environmental Research*, 70: 1–6.

Schulman, R., and Bucuvalas, Inc. (2001). *Confronting COPD in North America and Europe: A Survey of Patients and Doctors in Eight Countries.* New York.

Schwartz, J. (1994). Societal Benefits of Reducing Lead Exposure. *Environmental Research*, 66: 105–112.

Sengupta, I. (2007). Optimal Time for Investment to Regulate Emissions of Particulate Matter in Indian Thermal Power Plants. *Environment and Development Economics*, 12: 827–860.

Shanmugam, K. R. (1997). The Value of Life: Estimates from Indian Labour Market. *The Indian Economic Journal*, 44(4): 105–114.

Shi, A. (1999). The Impact of Access to Urban Potable Water and Sewerage Connection on Child Mortality: City Level Evidence, 1993. In S. Devarajan, F. H. Rogers, and L. Squire (Eds.), *World Bank Economists' Forum: Volume 1.* Washington, D.C.: World Bank.

Shibuya, K., Mathers, C., and Lopez, A. (2001). *Chronic Obstructive Pulmonary Disease (COPD): Consistent Estimates of Incidence, Prevalence, and Mortality by WHO Region* (Global Programme on Evidence for Health Policy). Geneva: World Health Organization,.

Shields, C. R., and Francois, J. F. (Eds.). (1994). *Modeling Trade Policy: Applied General Equilibrium Assessments of North American Free Trade*. New York: Cambridge University Press.

Simon, N., Cropper, M., Alberini, A., and Arora, A. (1999). *Valuing Mortality Reductions in India As Study of Compensating Wage Differentials* (World Bank Policy Paper). Washington, D.C.: World Bank.

Sinha, A. et al. (1999). Typhoid Fever in Children Aged Less Than Five Years. *Lancet*, 354: 734–737.

Strukova, E. (2004). *Opportunity Cost of Deforestation in the Brazilian Amazon: Aggregated Estimate Per Ton of Avoided Carbon Emission* (Working Paper). Washington, D.C.:World Bank.

Strukova, E. (2010). *Cost-Benefit Analysis of Development and Conservation Alternatives in the Sundarbans of India*. Climate Change Adaptation, Biodiversity Conservation & Socio-Economic Development of the Sundarbans Area of West Bengal World Bank Technical Assistance, unpublished. Washington, D.C.:World Bank.

Summers, L., and Zeckhauser, R. (2008). Policy Making for Posterity. *Journal of Risk and Uncertainty*, 37(2): 115–140.

Tata Energy Research Institute. (2001). *Review of Past and On-Going Work on Urban Air Quality in India* (TERI Project Report no. 2001 EE41). New Delhi: Tata Energy Research Institute.

———. (2003). *Electricity Externalities in India: Information Gaps and Research Agenda*. New Delhi: The Energy and Resources Institute.

ten Brink, P. (Ed.). (2011). *The Economics of Ecosystems and Biodiversity in National and International Policy Making*. London: Earthscan.

Thenkabail, P. S., Dheeravath, V., Biradar, C. M., Gangalakunta, O. P., Noojipady, P., Gurappa, C., Velpuri, M., Gumma, M., and Li, Y. (2009). Irrigated Area Maps and Statistics of India Using Remote Sensing and National Statistics. *Journal Remote Sensing*, 1: 50–67.

Tilman, D., and Downing, J. A. (1994). Biodiversity and Stability in Grasslands. *Nature*, 367: 363–365.

UN Department of Economic and Social Affairs, Population Division. (2009). *World Urbanization Prospects*. New York: United Nations.

UN Economic Commission for Europe. (2004). *Tajikistan Environmental Performance Review*. Available at: http://www.unece.org/env/epr/studies/Tajikistan/welcome.htm.

U.S. Environmental Protection Agency (EPA), Climate Change Division. (2008). The United States Environmental Protection Agency's Analysis of

Senate Bill S.2191 in the 110th Congress, the Lieberman-Warner Climate Security Act of 2008. Available at: http://www.epa.gov/climatechange/economics/economicanalyses.html.

Viscusi, W. K., and Aldi, J. E. (2002). *The Value of a Statistical Life: A Critical Review of Market Estimates Throughout the World* (Discussion Paper no. 392). Cambridge, MA: Harvard Law School.

Weyant, J. (Ed.). (1999). The Costs of the Kyoto Protocol: A Multi-Model Evaluation. *Energy Journal* (special issue).

WHO (2001). Quantification of health effects of exposure to air pollution. WHO Regional Office for Europe. Available at http://www.euro.who.int/document/e74256.pdf.

Wilson, D., and Purushothaman, R. (2003). *Dreaming with BRICS: The Path to 2050* (Global Economics Paper no. 99). New York: Goldman Sachs.

Washington, D.C.: World Bank. (2002). *Egypt Cost Assessment of Environmental Degradation (*Report no. 25175-EGT). Washington, D.C.: World Bank.

———. (2006). *Unlocking Opportunities for Forest-Dependent People in India* (Main Report, vol. 1. Report No. 34481-IN). Washington, D.C.: World Bank, Agriculture and Rural Development Sector Unit, South Asia Region.

———. (2007). *Cost of Pollution in China. Economic Estimates of Physical Damages*. Washington, D.C.: World Bank.

———. (2008). *World Development Indicators: Population Living in Slums*. Washington, D.C.: World Bank. Available at: http://data.worldbank.org/data-catalog/world-development-indicators.

———. (2010). *The Economic Impacts of Inadequate Sanitation in India*. Washington, D.C.: World Bank. Available at: http://www.wsp.org/wsp/sites/wsp.org/files/publications/wsp-esi-india.pdf.

———. (2011). *World Development Indicators 2011*. Washington, D.C.: World Bank.

———. (2012). *China 2030*. Washington, D.C.: World Bank.

World Health Organization. (2002a). *Global Burden of Disease 2002*. Geneva: World Health Organization.

———. (2002b). *The World Health Report 2002*. Geneva: World Health Organization.

———. (2004). *Global Burden of Disease 2004*. Geneva: World Health Organization.

———. (2008). *Air Quality and Health*. Available at: http://www.who.int/mediacentre/factsheets/fs313/en/.

Yang, H.-Y. (2001). Trade Liberalization and Pollution: A General Equilibrium Analysis of Carbon Dioxide Emissions in Taiwan. *Economic Modelling*, 18: 435–454.

Zamuda, C.D., and Sharpe, M.A. (2007). *A Case for Enhanced Use of Clean Coal in India: An Essential Step Towards Energy Security and Environmental Protection* (Workshop on Coal Beneficiation and Utilization of Rejects). Ranchi, India.Ministry of Food, Agriculture and Livestock (2006). Agricultural Statistics of India, Economic Wing, Ministry of Food, Agriculture and Livestock (2003–2004). New Delhi: Government of India.

Index